Underwater Intervention Conference (UI 2024)

New Orleans, Louisiana, USA
12-14 November 2024

ISBN: 979-8-3313-1283-1

Printed from e-media with permission by:

Curran Associates, Inc.
57 Morehouse Lane
Red Hook, NY 12571

Some format issues inherent in the e-media version may also appear in this print version.

Underwater Intervention
5206 Cypress Creek Parkway
Suite 202
Houston, Texas 77069
USA

Phone: 281-893-8539

www.underwaterintervention.com

Additional copies of this publication are available from:

Curran Associates, Inc.
57 Morehouse Lane
Red Hook, NY 12571 USA
Phone: 845-758-0400
Fax: 845-758-2633
Email: curran@proceedings.com
Web: www.proceedings.com

TABLE OF CONTENTS

OCEANEERING®

Sean Amenson and Katelyn Landry

Talent Acquisition Recruiters

We Solve the Unsolvable

It's a mindset, not a tagline. At Oceaneering, empowered individuals with unique and diverse perspectives accomplish great things. Our teams develop innovative technologies to solve critical challenges—from sea to space.

Oceaneering International, Inc.

In 1964, Mike Hughes and Johnny Johnson formed a Gulf of Mexico diving company called World Wide Divers, then in 1969 they merged with two other diving companies to form Oceaneering International, Inc. For over 60 years, the company has transformed from a small regional diving company into a global provider of engineered products and services.

60 Years of Experience in Harsh Environments

24 Operating Countries

70+ Operational Bases

12,000+ Valuable Employees

What We Do

Today, we develop products and services for use throughout the lifecycle of an offshore oilfield, from drilling to decommissioning. We operate the world's premier fleet of work class ROVs. Additionally, we are a leader in offshore oilfield maintenance services, umbilicals, subsea hardware, and tooling. We also serve the aerospace, defense, and theme park industries.

Industries we serve

- Aerospace
- Healthcare
- Logistics
- Manufacturing
- Themed Entertainment
- Defense
- Science and Research
- Energy

OCEANEERING®

Business Groups

Subsea Robotics (SSR)

Remotely Operated Vehicles (ROVs):
We operate the world's premier fleet of ROVs and are the leading provider of ROVs to the offshore energy industry.

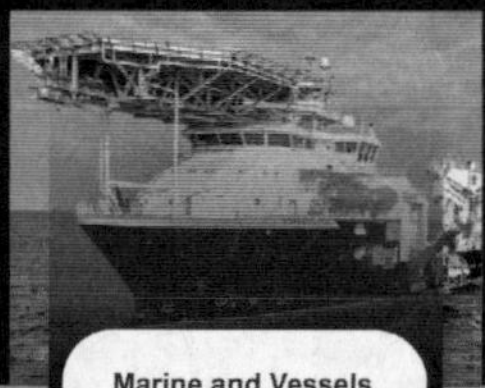

Marine and Vessels

Our integrated, complete vessel solutions combine remotely operated vehicle (ROV) systems, project management, installation engineering, advanced technologies, and offshore operational expertise to deliver a best-in-class solution.

Survey Services

Our survey and mapping services solve challenges in diverse applications, from offshore and onshore energy to marine construction and government operations.

Examples: AUV Pipeline Inspection Surveys, AUV Hazard Surveys, and Marine Construction Surveys

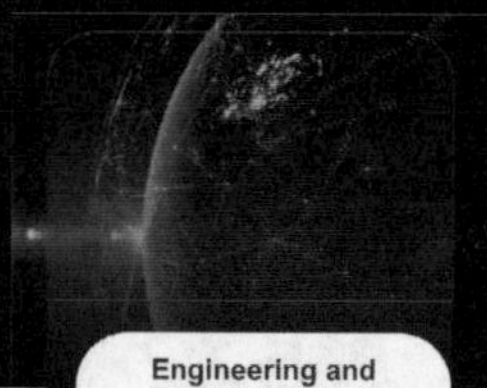

Engineering and Manufacturing

We have the experience and expertise to provide quality engineering and manufacturing of premier products.

Examples: Subsea Umbilicals, Subsea Distribution Systems, Entertainment Systems, ROVs

■ oceaneering.com/rov

OCEANEERING®

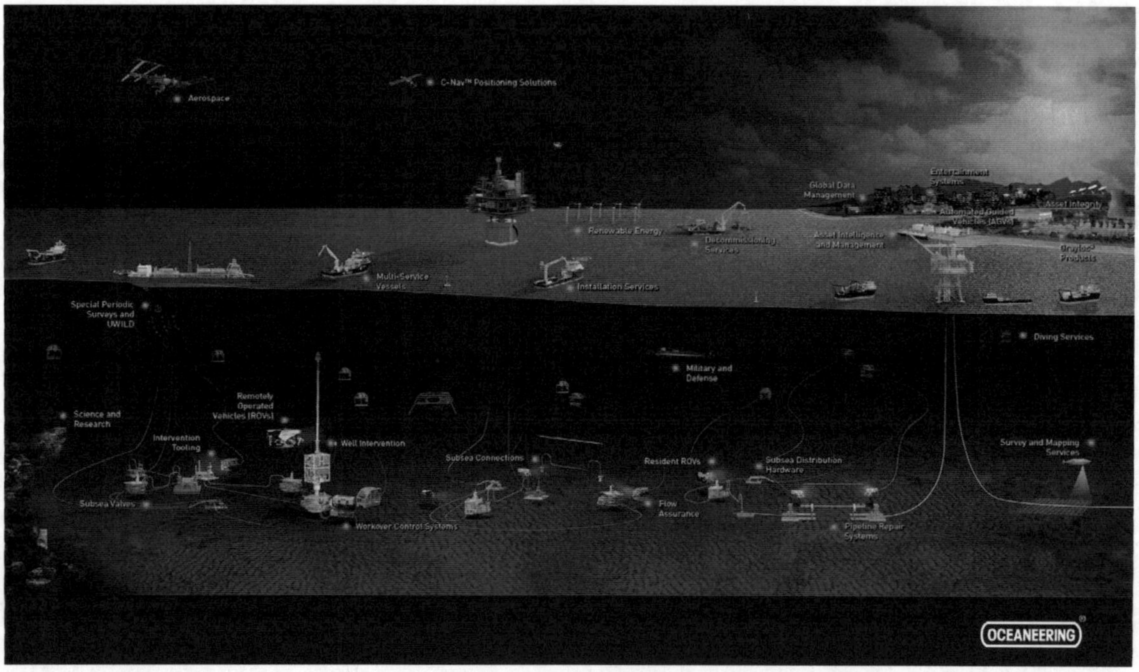

Common Jobs in Business Groups

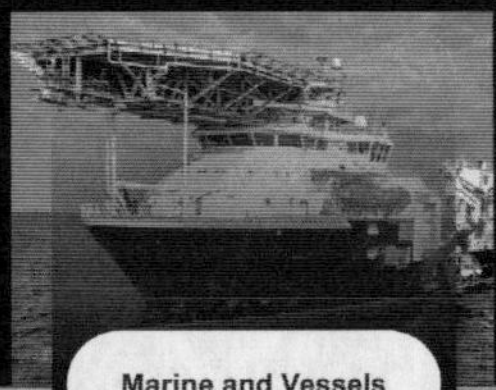

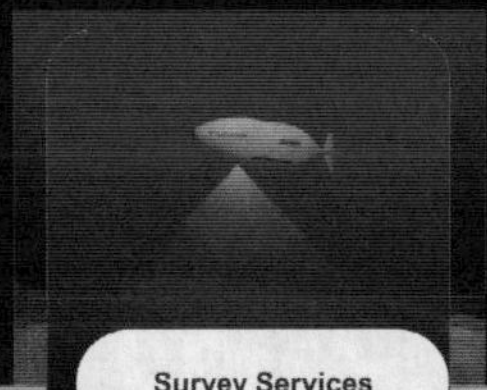

Subsea Robotics (SSR)

- Workshop Technician
- Subsea ROV Technicians
- Installation Technicians/Riggers
- Intervention Technicians

Marine and Vessels

- Vessel Galleyhand
- Vessel Cook
- Vessel QMED
- Vessel AB Seaman
- Vessel Bridge Officer
- Vessel Engineering Officer
- Vessel Electronics Technician
- Vessel Chief Engineering Officer
- Vessel Chief Officer
- Diving Tender

Survey Services

- Survey Equipment Technician
- Offshore Surveyors/Technicians
- Field Geologists
- Survey Data Processors

Engineering and Manufacturing

- Rotation Engineer
- Subsea Robotics Engineer
- Field Engineer
- Installation Engineer
- Lead Engineer
- Hardware Product Designer
- Machinist
- Quality Control Inspector/Engineer
- Assembly and Test Technician

■ oceaneering.com/rov

OCEANEERING

Internship Opportunities

- **All Internship opportunities should be posted on Oceaneering's Career site by January 15th, 2025.**
- **Follow Oceaneering on LinkedIn for 2025 Internship Updates.**

<u>2024 Internship Opportunities</u>

Manufacturing Engineering Intern	**Legal Intern**
Electrical Engineer Intern	**SCM Logistics Intern**
Software Engineering Intern	**IT Intern - Integration Developer**
Cybersecurity Intern	**Contracts Analyst Intern**
QA Tester - IT Applications Intern	**Commercial and Sales Intern**
HR Intern	**Business Intelligence Analysis Intern**
Project Coordinator Intern	**HSE Intern**

OCEANEERING

Oceaneering Career Site

1. Go to Oceaneering.com
2. Click on Careers tab
3. Click Career Search
4. Choose which region to search for open positions in and search opportunities

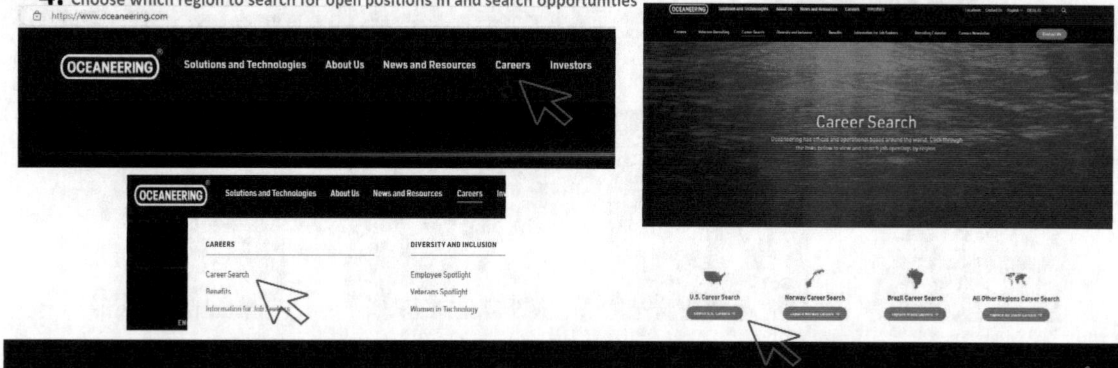

OCEANEERING

Stay in Touch and Follow Us!

Follow us on our social media platforms to stay informed and up-to-date with employment opportunities.

@Oceaneering

OCEANEERING

Questions?

OCEANEERING®

Connecting What's Needed with What's Next™

Peerside:
Ocean STEAM Workforce Development

Colbi Gemmell
William 'Monty' Graham
Florida Institute of Oceanography

PEERSIDE℠

Peerside Program

Peerside aims to broaden access and ongoing involvement with the ocean.

A cohort approach allows exposure to varied career opportunities and provides a chance for peer interaction, adequate mentorship and network building.

PEERSIDE℠

2

Peerside Program: Stemmed in Partnership

Peerside builds on ocean community approach by strengthening and developing partnerships with various organizations:

- Support the missions of partner organizations
- Build the blue economy workforce
- Increase ocean inclusion
- Expand programmatic reach

PEERSIDE℠

3

The PEERSIDE℠ Model

- Mentees: Undergrads from 2-year and 4-year institutions, grad students, early career post-grads
  - Engage in:
    - Professional development, experiential learning and remote science opportunities
    - Sustained community and mentorship through career stages
    - Skill development for participation in the blue economy
- Mentors: Various ocean science professionals committed to providing guidance through mentorship
  - Provided:
    - Mentorship training
    - Chance to develop new science and technical skills
    - Expedition planning and research opportunities
    - Professional development
    - Networking

4

Learning Opportunities and Professional Development

- Participants engage in monthly webinars
- Mentees work with their mentors at their host institutions/organizations

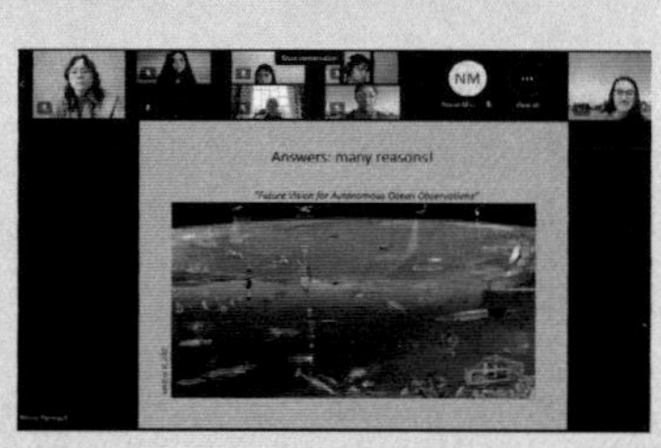

PEERSIDE℠

5

Dedicated Program Vessel: *R/V Western Flyer*

- 117-foot swath vessel
- Berthing for 16 science and ROV personnel
- Moon pool for ROV deployment
- ROV control room

PEERSIDE℠

6

Program Tool: ROV Taurus

- Brought on board in May 2024
- Designed by Pelagic Research Services and funded through ONR
- Depth rated to 4000m

PEERSIDE℠

7

ROV Taurus and Peerside

The ROV provides interdisciplinary deep sea research opportunities

- Mixing science, technology, social science and art to obtain research outcomes
- Sample and tooling opportunities
- Allows for ship to shore connections
- Provides necessary blue economy skills

PEERSIDE℠

8

ROV Blue Economy Applications

- BOEM mineral exploration
- Deep sea organism biopharmaceutical connections
- Seafloor mapping
- Archaeology

PEERSIDE℠

9

ROV Tooling and Operating Experience

- Participants from Long Beach City College were able to design and apply ROV tools based on science objectives
- Provided opportunities to work with Pelagic Research Services at sea, deploying instrumentation and getting ROV piloting opportunities

PEERSIDE℠

10

Remote Science Opportunities: Connecting Ship to Shore

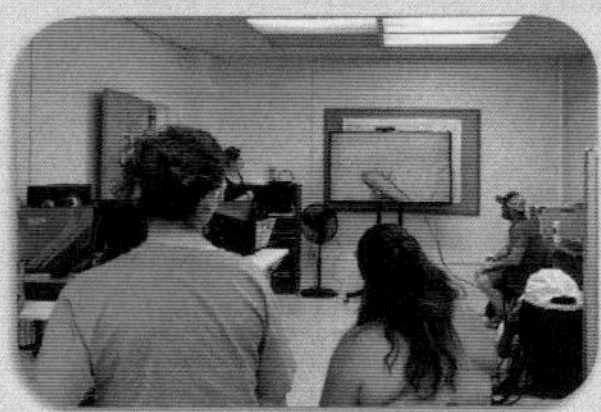

- Participants co-direct at sea expeditions on-shore in real time
- Utilizes telepresence through associated technologies
- Shore team involved in the expedition planning and execution process

PEERSIDE℠

11

Experiential Learning Opportunities

Cohorts participate in the following activities:

- 6 day at-sea expedition
- Community-based remote science
- Field trips

PEERSIDE℠

12

Connecting with Ocean Professionals and Networking

In addition to working with mentors throughout the program, participants connection with other ocean professionals and peers, expanding their ocean community

– Participate in career talks
– Field trips to partner organizations

PEERSIDE℠

13

Program Feedback and Benefits

– Peerside allows for a sense of belonging within the ocean community.
– Increased knowledge and confidence by both mentors and mentees
– Exposure to diverse interdisciplinary career opportunities

PEERSIDE℠

14

Next steps and Future Directions

- Peerside is a tiered process with FIO serving as the model iteration
- Further engagement with maritime professional trades and US defense networks
- Greater connection to the K-12 community

15

Want to get involved?

- Potential participants can submit an Expression of Interest through our website: https://www.fio.usf.edu/peerside/
- You can email us at fio-peerside@usf.edu

PEERSIDE℠

16

Thank You!
Office of Naval Research
ONR
Science & Technology
SCHMIDT
OCEAN
INSTITUTE
SCHMIDT FAMILY
FOUNDATION
NSF
PEERSIDE
17

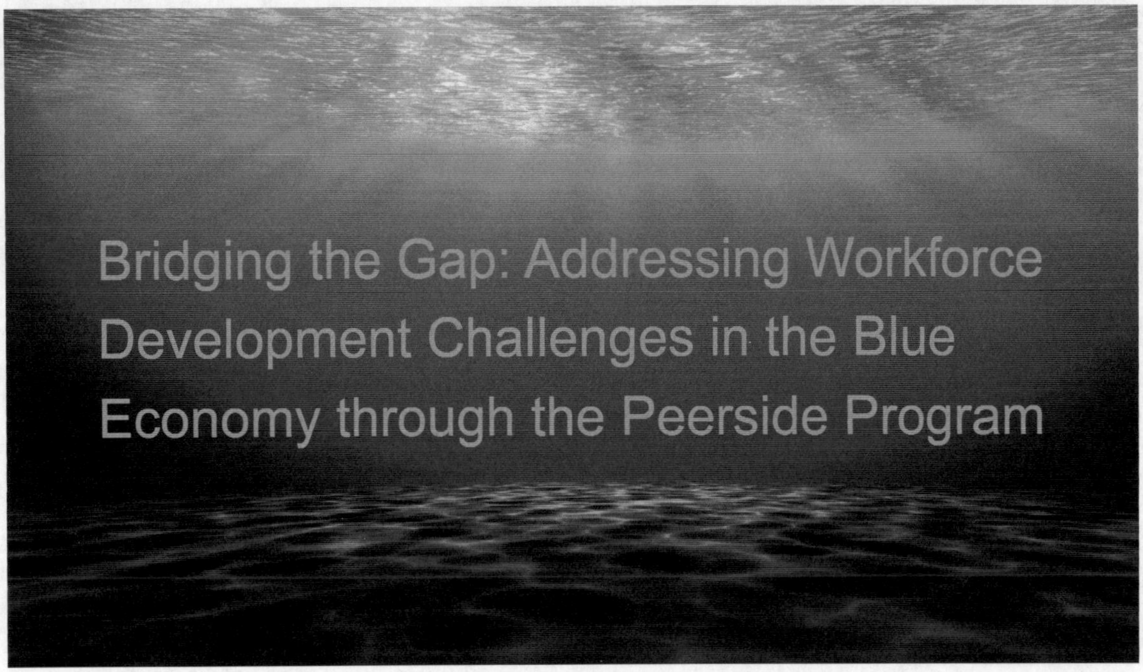
Bridging the Gap: Addressing Workforce Development Challenges in the Blue Economy through the Peerside Program

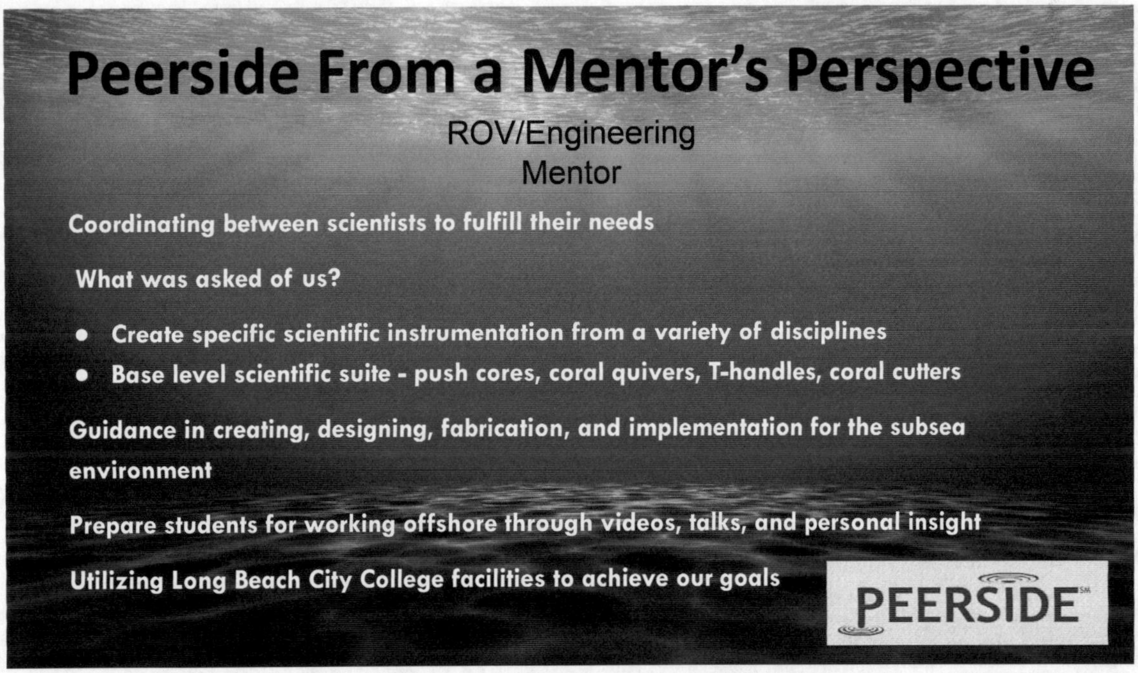
Peerside From a Mentor's Perspective
ROV/Engineering
Mentor
Coordinating between scientists to fulfill their needs
What was asked of us?
Create specific scientific instrumentation from a variety of disciplines
Base level scientific suite - push cores, coral quivers, T-handles, coral cutters
Guidance in creating, designing, fabrication, and implementation for the subsea environment
Prepare students for working offshore through videos, talks, and personal insight
Utilizing Long Beach City College facilities to achieve our goals
PEERSIDE

Joseph - What I was Hoping For

- Working conceptualize tooling for scientists
- Design hardware that's easy to implement and conducive to science
- Hands-on experience implementing hardware onto an existing ROV

PEERSIDE℠

Projects & Experience

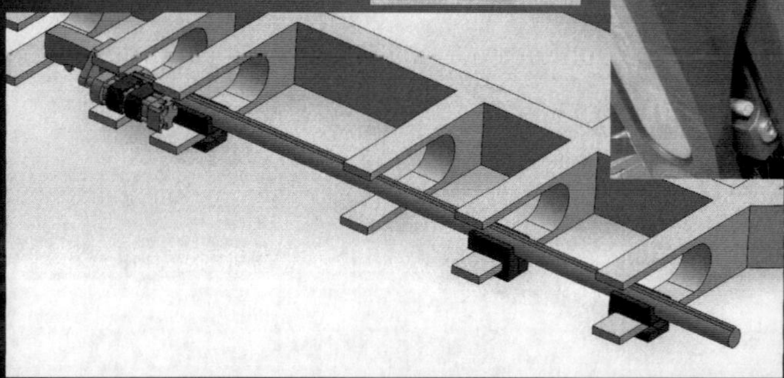

PEERSIDE℠

Projects & Experience
PEERSIDE℠

Projects & Experience
PEERSIDE℠

Projects & Experience

PEERSIDE℠

Peerside From a Mentor's Perspective

This opportunity provided me with countless beneficial experiences & exposure to key

Exposure - wide range of professions and possibilities within the industry

Education - Industry overviews, career discussions, professional development modules, etc.

Mentorship - Alignment with mentors, not only in our discipline but other disciplines as well

Network - This experience has introduced me to many professionals that have opened up additional opportunities

PEERSIDE℠

Access to Software and Design Engineers, Survey team, ROV Pilots & Technicians

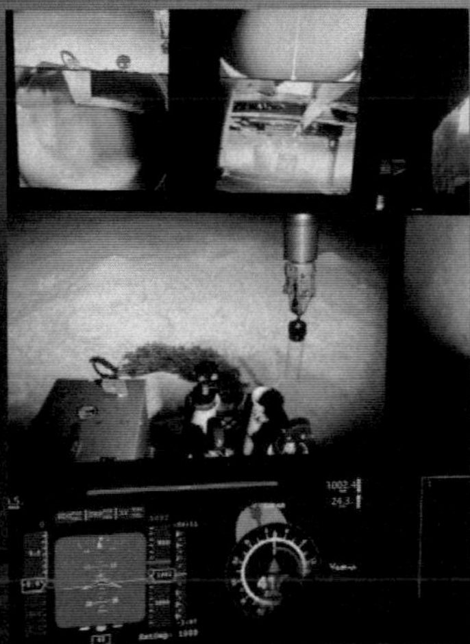

Favorite Part: Supporting Exploration of Parts of the World not often Frequented

Long Beach city College (LBCC) Produces Talent Ready for the Workforce

What do we have to offer?

- Numerous opportunities within our technical trade programs.
  - Electrical trade program E= I x R
  - Motor control classes to understand how we can control motor speeds to improve our efficiency
  - Programmable logic control (PLC) function. How stuff works when set with controls
  - Generator principles and function
  - Robotics how to program a system to perform as desired.
  - Eager students ready to hit the workforce and apply their training.

How do we do this?

- Backing up lectures with practical applications in our labs to reinforce their training.
- Working on projects to gain experience in different areas of training like high voltage, robotics and designing things in Computer aided drafting.
- Seeing their ideas through to completion, come to life if you will.

We support this by

- Having the necessary expertise and support to allow innovation and creativity to flourish.
- Having the tools to build

- 3d printing creating smaller practical items for everyday uses and advancements in science

How does this prepare them for their future ?

- Working on items relevant to the needs of the blue economy.
- LBCC currently has a student driven project to build an ROV simulator with a control van to prepare them for this type of work and to become familiar with expectations

How does this benefit you?

- You get an individual better prepared to meet challenges of the future.
- Untapped resource of eager people that had no idea this avenue was available to them.
- Eager people that are ready to contribute their talent to any company that will offer them an opportunity.
- Look no further than to two of of our speakers Jason and Joseph excellent examples of the LBCC student talent pool.

BlueRobotics
We're on a mission to enable the future of marine robotics

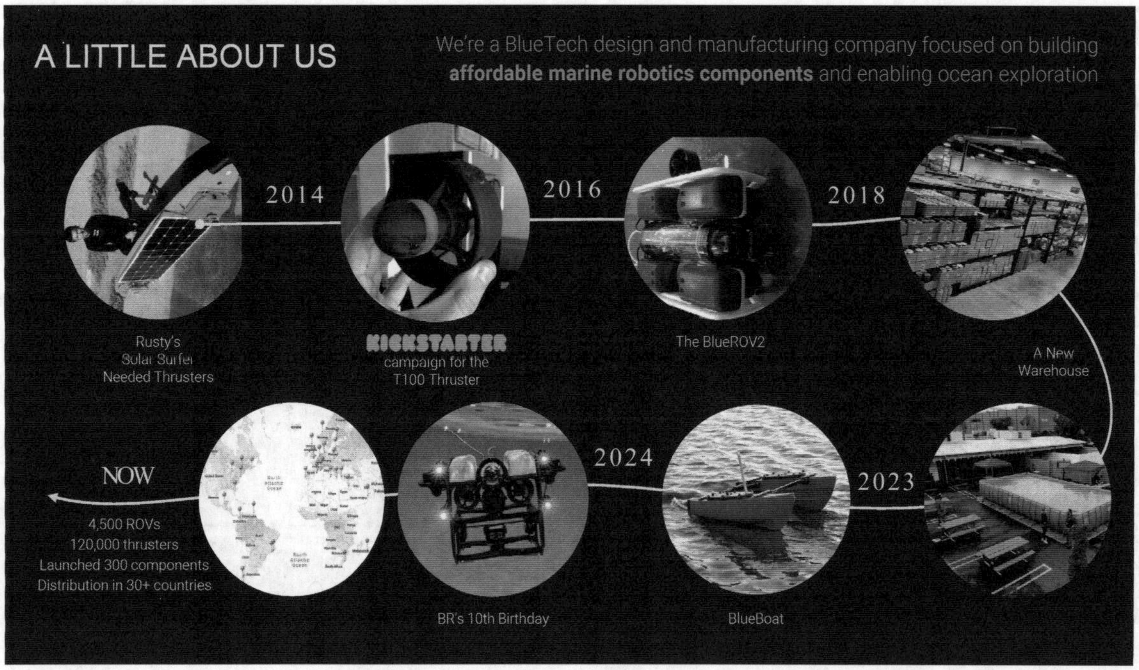
A LITTLE ABOUT US
We're a BlueTech design and manufacturing company focused on building **affordable marine robotics components** and enabling ocean exploration
2014
2016
2018
Rusty's Solar Surfer Needed Thrusters
KICKSTARTER campaign for the T100 Thruster
The BlueROV2
A New Warehouse
NOW
2024
2023
4,500 ROVs
120,000 thrusters
Launched 300 components
Distribution in 30+ countries
BR's 10th Birthday
BlueBoat

WHY YOU SHOULD KNOW ABOUT MARINE ROBOTICS

INDUSTRIES

- Education
- Exploration
- Environmental Management
- Infrastructure Services
- Military / Defense
- Science / Research / Conservation
- First Responders / Police / Fire + Rescue

APPLICATIONS

- Aquaculture, monitoring, + sampling
- Energy Production + Distribution
- Inspection + Observation
- Telecommunication
- Video + Photography
- Recreation
- Search + Rescue (SAR)

The BlueROV2 with NASA's ROV, Bruie, designed to conduct under-ice research so we can someday compare our oceans to the frozen oceans of Jupiter's moon, Europa! This will help determine if there might have been life on Europa.

HOW WE SUPPORT EDUCATION

SUPPLY COMPONENTS

K-12, Junior Colleges, Colleges & Universities

EDUCATIONAL DISCOUNT CODE: "supportstem"

SPONSOR COMPETITIONS

MATE, RoboNation, SAUVC, and many more!

WHERE TO GO FROM HERE

Education & Career Opportunities:

ENGINEERING

- Mechanical / CAD
- Electrical / PCB Design
- Software / Coding
- Manufacturing

SALES + MARKETING

- Technical Sales
- Marketing / Promotion
- Community / Fundraising
- Communication / Writing
- Creative Arts

SCIENCES

- Marine Sciences + Research
- Environmental Studies + Conservation
- Mechatronics

OTHER

- ROV Technician
- ROV Pilot
- Entrepreneurship (SAR, Inspection)
- Technical Support

Santa Cruz Island Field Trip -
Kayak Deployment
Marine Biology Class,
Channel Islands High School,
Oxnard, CA

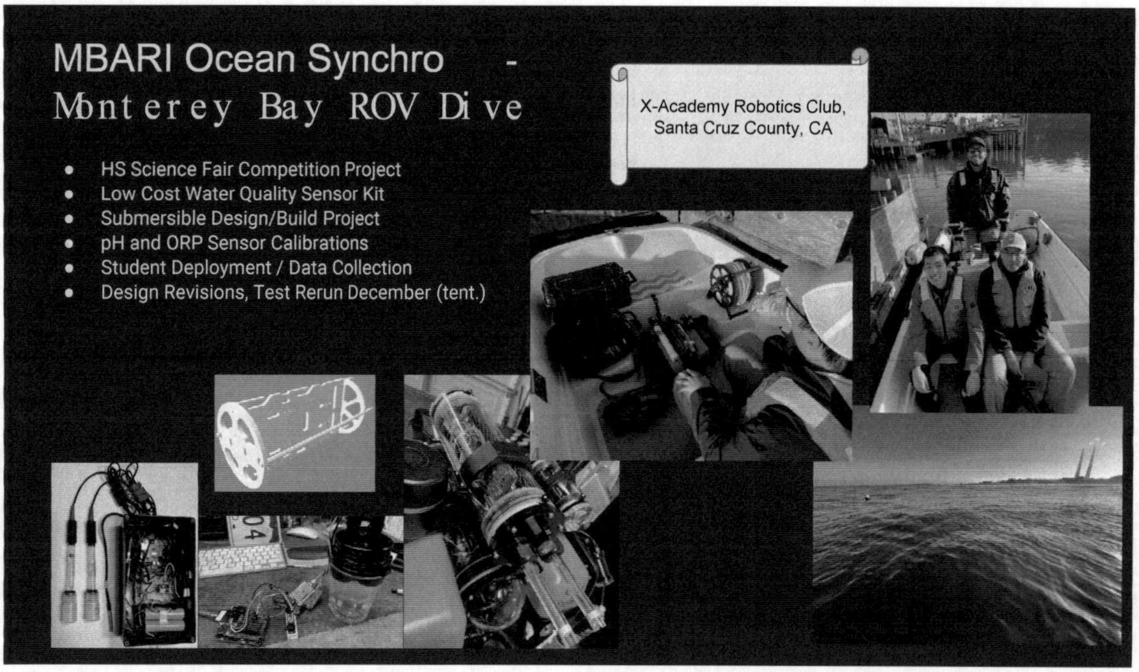
MBARI Ocean Synchro -
Monterey Bay ROV Dive
HS Science Fair Competition Project
Low Cost Water Quality Sensor Kit
Submersible Design/Build Project
pH and ORP Sensor Calibrations
Student Deployment / Data Collection
Design Revisions, Test Rerun December (tent.)
X-Academy Robotics Club,
Santa Cruz County, CA

Pool Practice & Anacapa
Island ROV Dive
Environmental Studies Class
Buena High School,
Ventura, CA

sea
ya
BlueRobotics

Uncrewed Multi-Payload Operations with a High-Power Long-Endurance USV
Michael P. Scherer, CTO
Michael@ChanceMaritime.com
Stuart Chance, CEO
Stuart@ChanceMaritime.com
UNCREWED BOATS
DESIGN · BUILD
OPERATE · LEASE
Business Confidential / Proprietary Information
Copyright © 2024 Chance Technologies, LLC

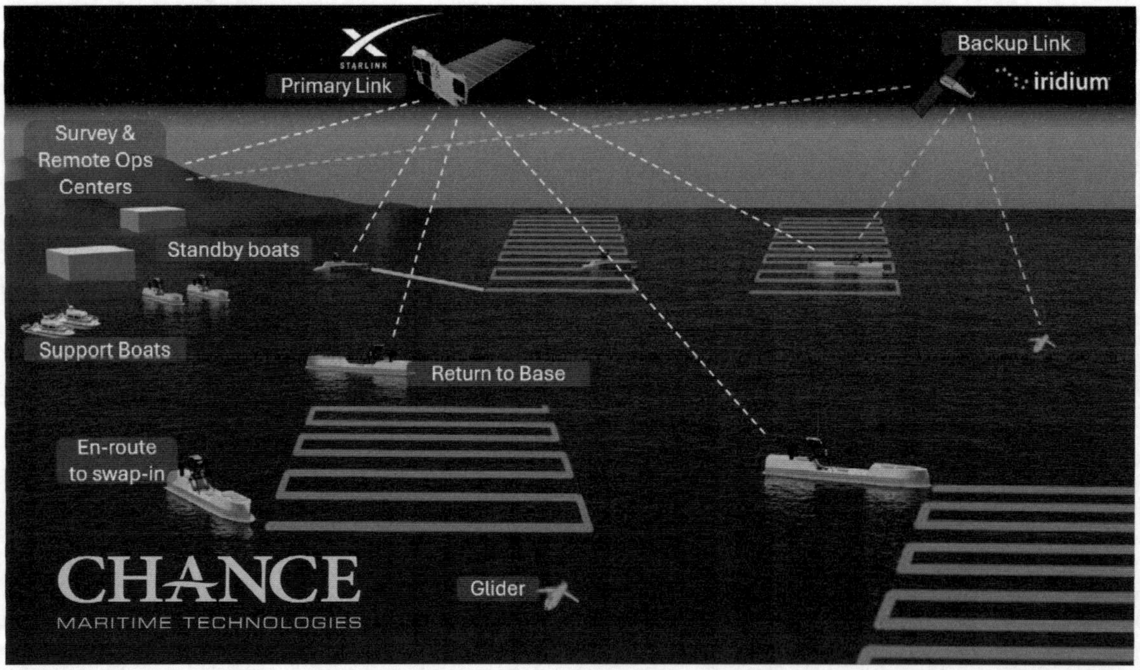
STARLINK
Primary Link
Backup Link
iridium
Survey & Remote Ops Centers
Standby boats
Support Boats
Return to Base
En-route to swap-in
Glider
CHANCE
MARITIME TECHNOLOGIES

Survey &
Remote Ops
Centers
Standby
CHANCE
MARITIME TECHNOLOGIES

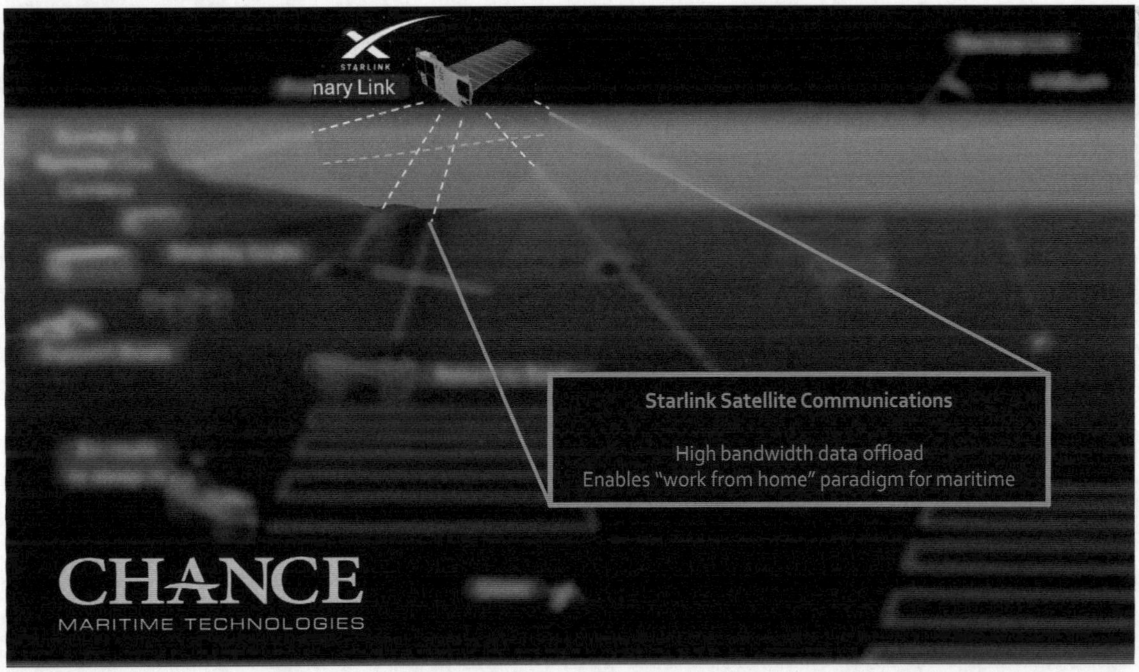
STARLINK
nary Link
Starlink Satellite Communications
High bandwidth data offload
Enables "work from home" paradigm for maritime
CHANCE
MARITIME TECHNOLOGIES

MC-29	
Dimensions	29.0 ft x 5.8 ft [8.84 m x 1.77 m]
Draft	2.8 ft [0.85 m]
Top Speed	7.0 kts
Cruise Speed	4.5 kts
Endurance	20+ days at Cruise Speed
Maximum Range	2150 nm
Payload Deployment	Moon Pool, Deck Mount, A-Frames
Unique Features	Redundant Propulsion

CHANCE
MARITIME TECHNOLOGIES

Planning is Key!

CHANCE
MARITIME TECHNOLOGIES

- Understanding the operational environment:
  - Access, Egress, Navigation Hazards, Weather all important factors
- Logistics can make or break your operations:
  - Our USV's are road trailerable, and easy to deploy down a slipway
  - Long endurance vehicles reduce launch and recovery costs
- Integration of sensors:
  - Ability to drill into the hull to mount anything
  - Moon pool, A-frames, winches, 19" rack
  - Climate controlled spaces
  - Available diesel for auxiliary payloads

Case Study: CPRA

CHANCE
MARITIME TECHNOLOGIES

- Customer:
  - HydroTerra Technologies for the Louisiana Coastal Protection & Restoration Authority (CPRA)
- Mission:
  - Bathymetric data collection for 130,000 acre region in Southwest Louisiana
- Auxiliary Payloads:
  - Multiparameter Sonde from ULL LO-SPAT to collect biological data to support Oyster Fisheries

Read More >

7

CPRA Survey Operations

CHANCE
MARITIME TECHNOLOGIES

- Navigation Hazards
  - Shallow waters, down to 3ft
  - Rigs with dubious charting and/or lighting
  - Shrimp boats
- Mitigations:
  - Shallow draft USV with protected running gear
  - Bring in latest BSEE data
  - Active radar, 360 visual, thermal cameras, Class A AIS, remotely operable VHF radio

Results

CHANCE
MARITIME TECHNOLOGIES

- Operated the MC-29 from over 100 miles away using Starlink
- Completed 670 Nautical Miles of bathymetric survey for CPRA and HydroTerra on half a tank
- 8 days to complete survey block with 24 hour operations

Key Findings

Area was too shallow for a liveaboard to do 24/7 operations.

MC-29 is unique for both its long endurance and shallow draft that allowed it to complete this large block in a short period of time

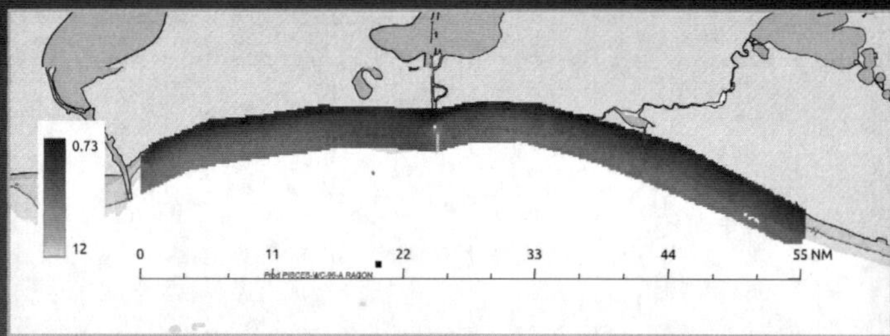

Multibeam Bathymetry

CHANCE
MARITIME TECHNOLOGIES

- Inland & Offshore Data Collection
- Winch system onboard used for remote sound speed profiling
- System mobilized and patch tested within two days

Multibeam	R2Sonic 2022 V+
IMU (Integrated)	SBG Pulse40-43T
Sound Speed	Valeport Swift SVP
Acquisition Software	Hypack Hysweep R2Sonic Control 2023

Results

CHANCE MARITIME TECHNOLOGIES

- Processing conducted with
  - Qinertia for ephemeris position corrections
  - Hypack Hysweep Editor – Multibeam Editor
- Excellent data quality, 1x1 ft grid

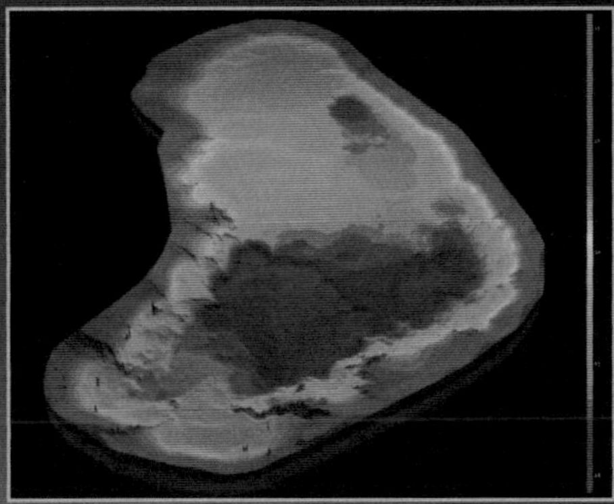

Key Findings

The MC-29 provided ample power to continuously operate the multibeam.

Active roll stabilization from the onboard SeaKeeper mitigated any roll effects while operating in Vermilion Bay

ROV Integration

iCORE CHANCE MARITIME TECHNOLOGIES

- Integrated Inspection Class ROV to MC-29 USV
- Collaboration with LSU iCORE Lab / Dr. Corina Barbalata
- Major Systems:
  - Launch & Recovery
  - ROV Command & Control
  - ROV Video

Remote Operations
ROV & USV operators side-by-side to coordinate operations
CHANCE
MARITIME TECHNOLOGIES

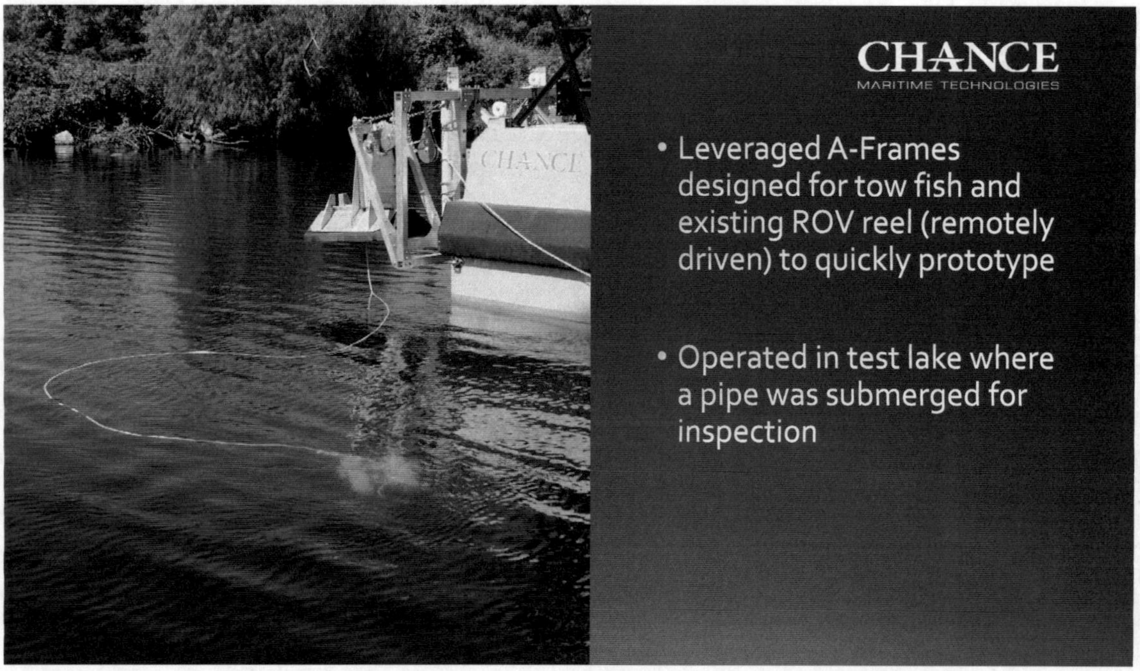
CHANCE
MARITIME TECHNOLOGIES
• Leveraged A-Frames designed for tow fish and existing ROV reel (remotely driven) to quickly prototype
• Operated in test lake where a pipe was submerged for inspection

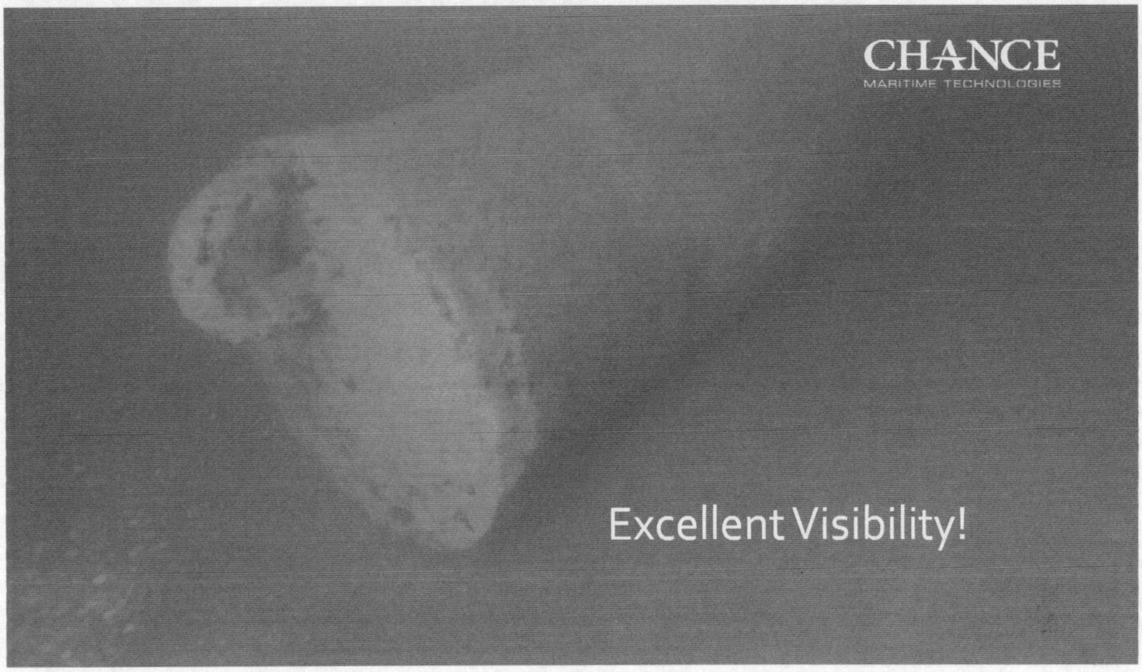
CHANCE
MARITIME TECHNOLOGIES
Excellent Visibility!

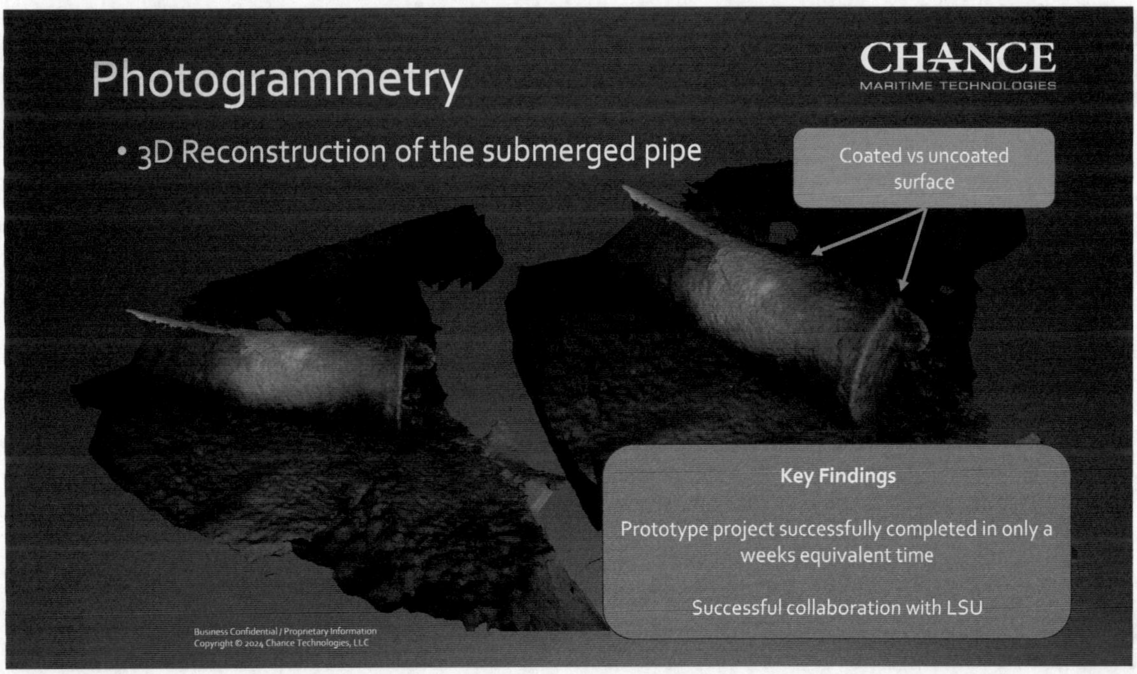
Photogrammetry
CHANCE
MARITIME TECHNOLOGIES
• 3D Reconstruction of the submerged pipe
Coated vs uncoated surface
Key Findings
Prototype project successfully completed in only a weeks equivalent time
Successful collaboration with LSU
Business Confidential / Proprietary Information
Copyright © 2024 Chance Technologies, LLC

Future Operations

CHANCE
MARITIME TECHNOLOGIES

- Long Endurance MC-40, projected 90 day endurance
  - Transit from San Diego to Japan and back again on a single tank
- Evolve ROV integration
  - Larger ROV's and open ocean capable LARS
- New class LR-30 initial sea trials

CHANCE
MARITIME TECHNOLOGIES

UNCREWED BOATS
DESIGN · BUILD
OPERATE · LEASE

www.ChanceMaritime.com

Business Confidential / Proprietary Information
Copyright © 2024 Chance Technologies, LLC

Chance Maritime Technologies

CHANCE
MARITIME TECHNOLOGIES

USVs Leased as a Service

- Founded in 2022 by former ASV Global / L3Harris employees
- 15 Full Time Employees and Growing
- 120+ years of USV experience
  - +75 USVs ranging from 1m to 100m in length
  - +30 payload integrations
- **Chance Maritime:**
  - **Designs, Builds, and Operates USVs**
  - **Integrates Client's payload into USV**
  - **Operates the USV while relaying data back to Client**
- Facilities Include:
  - Remote Operations Center
  - Climate-Controlled Production Facility
  - Large Yard
  - Private Test Lake (22 acres, 15.2 m deep)
- Industrial partnerships with University of New Hampshire Center for Coastal and Ocean Mapping Joint Hydrographic Center (CCOM/JHC) and COMIT / USF

19

Unique Uncrewed Boats

CHANCE
MARITIME TECHNOLOGIES

- Heavy Aluminum Construction
- Reliable Diesel Propulsion, Low Carbon Footprint
- Industry Leading Power + Endurance
- Easy Logistics / Road Trailerable
- Active Gyrostabilization
- Starlink Connectivity

✓ Payload Agnostic

✓ Kilowatts of Power

✓ Weeks of Endurance

20

Mission Capable 29 (MC-29)

CHANCE
MARITIME TECHNOLOGIES

Long endurance multipurpose platform for exploration inshore or offshore

- "Pickup-truck" functionality to support variety of mission payloads and deployment requirements
- Deployment through large moonpool or from deck
- 1-month long endurance capability
- Integrated gyrostabilizer for maximum stability
- Standard trailer capable for maximum logistic flexibility

MC-29		
Dimensions	29.0 ft x 5.8 ft	[8.84 m x 1.77 m]
Draft	2.8 ft	[0.85 m]
Top Speed	7.0 kts	
Cruise Speed	4.5 kts	
Endurance	20+ days at Cruise Speed	
Maximum Range	2150 nm	
Payload Deployment	Moon Pool, Deck Mount, A-Frames	
Unique Features	Redundant Propulsion	

21

Platforms – MC-40

CHANCE
MARITIME TECHNOLOGIES

Persistent High-Power + Long-Endurance

- Direct drive diesel propulsion
- Backup diesel generator & retractable azimuthing drive
- Flexible moon pool and mission bay space
- Robust ¼" aluminum construction
- Integrated gyrostabilizer for maximum stability
- Trailerable, 40' FEU container transportable

MC-40		
Dimensions	39.2 ft x 5.8 ft	[11.95 m x 1.77 m]
Draft	4.3 ft	[1.31 m]
Top Speed	7.0 kts	
Cruise Speed	4.5 kts	
Endurance	90+ days at Cruise Speed	
Maximum Range	9700 nm	
Payload Deployment	Moon Pool, Astern Mission Bay	
Unique Features	Onboard Generator	

90+ Day Endurance

22

Platforms – Line Runner 30 (LR30)

CHANCE
MARITIME TECHNOLOGIES

Long endurance hydrographic vessel

- Optimized for hydrographic survey operations
- Wave piercing bow for long endurance; est. 14 days at survey speed
- Integrated gyrostabilizer for maximum stability
- Standard trailer capable for maximum logistic flexibility
- Rails on deck for configurable payloads, A-frames, winches

LR-30		
Dimensions	30.0 ft x 6.5 ft	[9.14 m x 1.98 m]
Draft	3.0 ft	[0.91 m]
Top Speed	12.0 kts	
Cruise Speed	7.0 kts	
Endurance	14+ days at Cruise Speed	
Maximum Range	2150 nm	
Payload Deployment	Modular Keel & Deck Mounts, A-Frame	
Unique Features	Optional Davit Deployment	

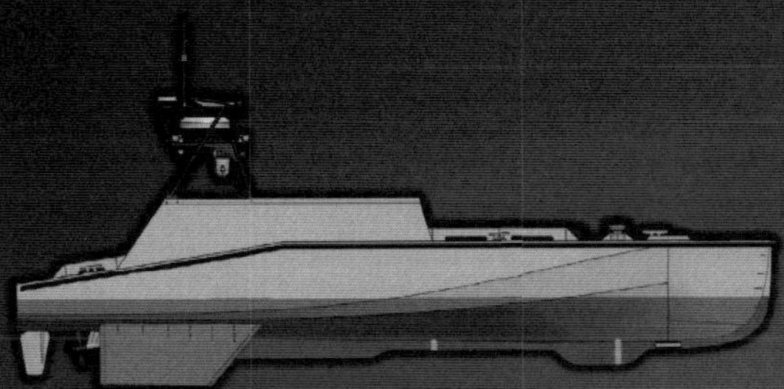

23

Classic Force Multiplication

CHANCE
MARITIME TECHNOLOGIES

- Our staff pioneered force multiplication with the first data from a USV integrated on NOAA charts – operational location was in Alaska
- Shown to increase operational efficiency by > 40%
- Can use multiple USV's

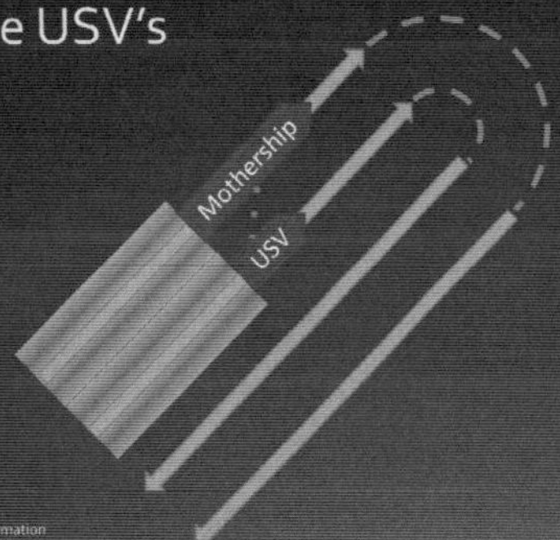

ASV working alongside Q105 in the Bering Strait. 2015.

Orthmann, Andrew, and Andrew Ziegwied. "The force multiplier effect using autonomous surface vessels for hydrographic survey in the arctic." OCEANS 2017-Anchorage. IEEE, 2017.

24

OTH Force Multiplication

CHANCE
MARITIME TECHNOLOGIES

- Enabled by Starlink™ satellite communication, the USV can now work completely independent of the Mothership
- Requires longer endurance USV's, such as the LR-30, that can operate for at least a week at a time

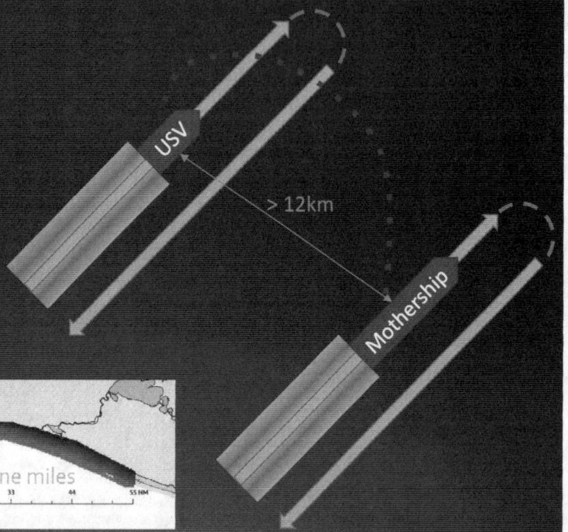

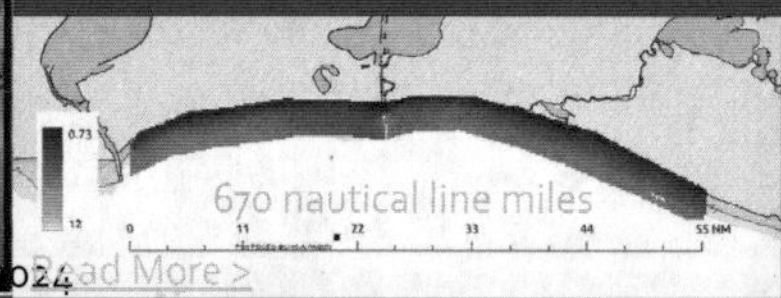

Over-The-Horizon work of MC-29 in 2024 Read More >

25

Questions?

CHANCE
MARITIME TECHNOLOGIES

Thank you on behalf of the rapidly growing Chance Maritime Team!

For additional questions:
Stuart Chance | Stuart@ChanceMaritime.com | 337-278-7746
Michael Scherer | Michael@ChanceMaritime.com | 727-743-0451
David Comeaux | David.Comeaux@ChanceMaritime.com | 337-258-3831

26

MTS
marine technology society
Marine
Technology
Society
MICROCREDENTIALS

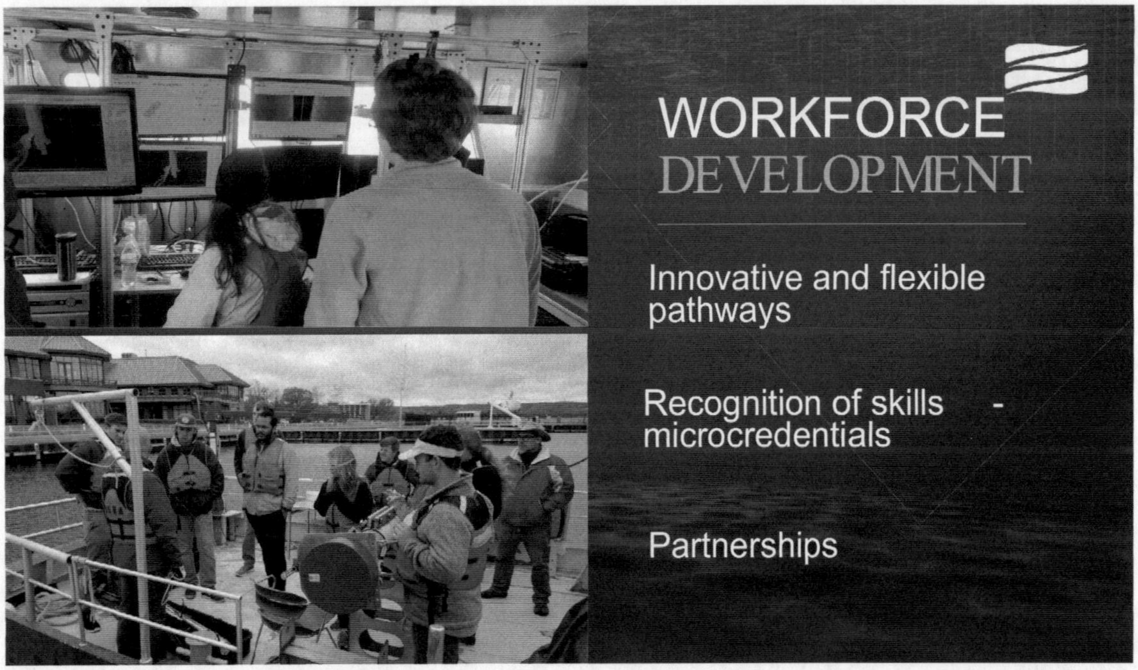
WORKFORCE
DEVELOPMENT
Innovative and flexible pathways
Recognition of skills - microcredentials
Partnerships

MicroCREDENTIALS

Ability Levels	
Foundational (Learning focus)	Validated Experience: • Fundamentals of the technology • Beginning concepts • Uses/application • Basic knowledge demonstrated
Intermediate (Learning and competency demonstration/practicum)	Validated Experience: • Technology applications • Intermediate concepts • More complex uses • Intermediate level knowledge demonstrated • PD at work and application
Advanced (Competency mastery through demonstration)	Validated Experience: • Practical experience (months to years) • Capstone Experiences – demonstration of abilities O Platforms O Sensors O Data

Building an Ocean Technician Workforce

The most effective preparation combines several different types of training, to ensure that technicians have the opportunity to learn in a variety of ways and master the skills they need to be successful in their role.

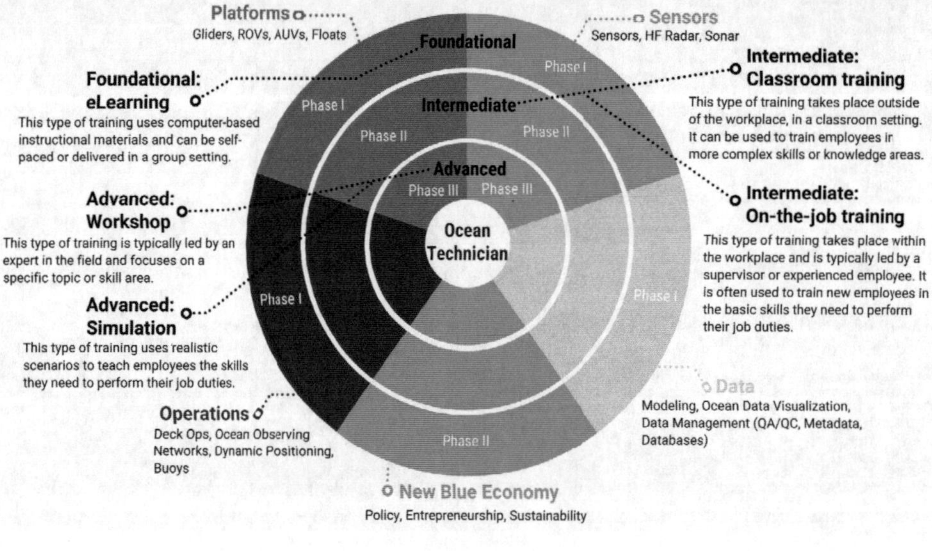

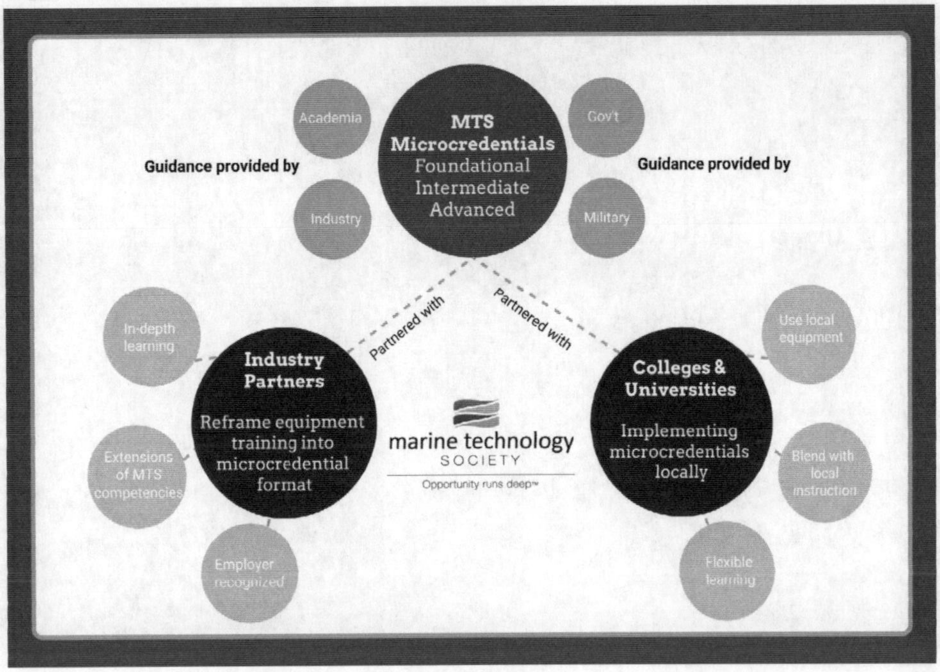
MTS
Microcredentials
Foundational
Intermediate
Advanced
Academia
Industry
Gov't
Military
Guidance provided by
Guidance provided by
Partnered with
Partnered with
Industry Partners
Reframe equipment training into microcredential format
In-depth learning
Extensions of MTS competencies
Employer recognized
marine technology SOCIETY
Opportunity runs deep™
Colleges & Universities
Implementing microcredentials locally
Use local equipment
Blend with local instruction
Flexible learning

MTS
marine technology society
Opportunity runs deep™
mtsociety.org

Sonardyne
Enabling subsea launch, recovery and residency with acoustic and optical positioning, navigation and comms
Aidan Thorn
Business Development Manager – Marine Robotics
© Sonardyne International | Commercial in Confidence

Subsea Robotics: Increasingly capable systems
Sonardyne
© Sonardyne International | Commercial in Confidence

Marine Robotics: Comms is key - Acoustics

Sonardyne

Marine Robotics: Comms is key - Optical

Sonardyne

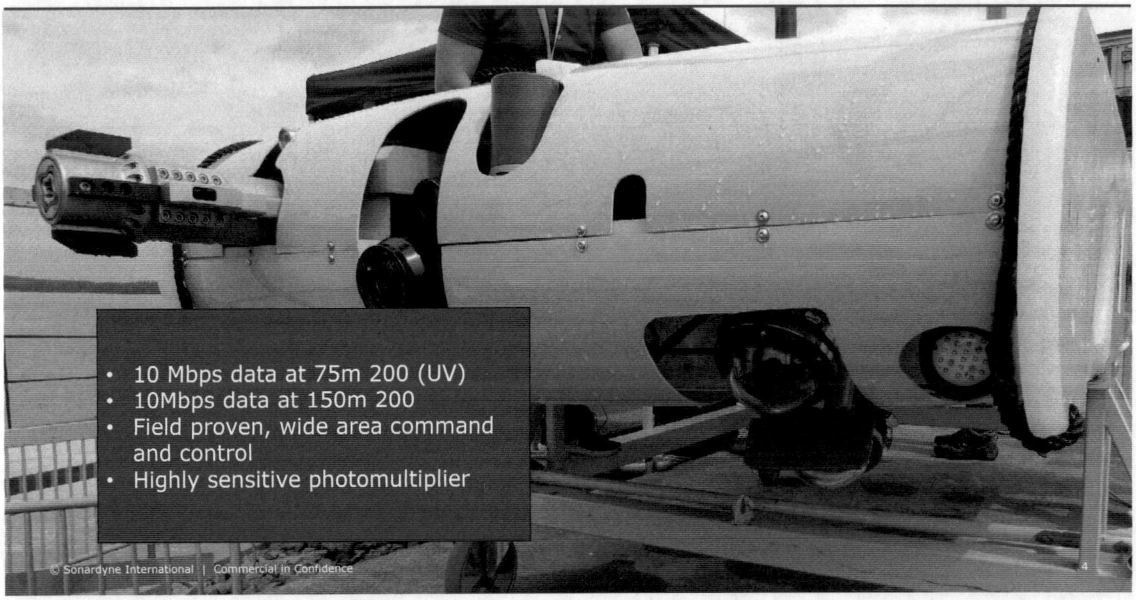

- 10 Mbps data at 75m 200 (UV)
- 10Mbps data at 150m 200
- Field proven, wide area command and control
- Highly sensitive photomultiplier

Marine Robotics: Positioning – LBL and SPRINT-Nav

Sonardyne

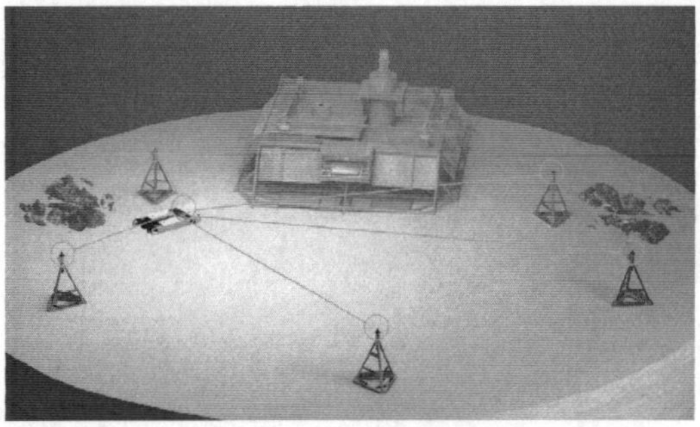

Sonardyne

Case Studies

Marine Robotics: Field Residency Proof of Concept

Sonardyne

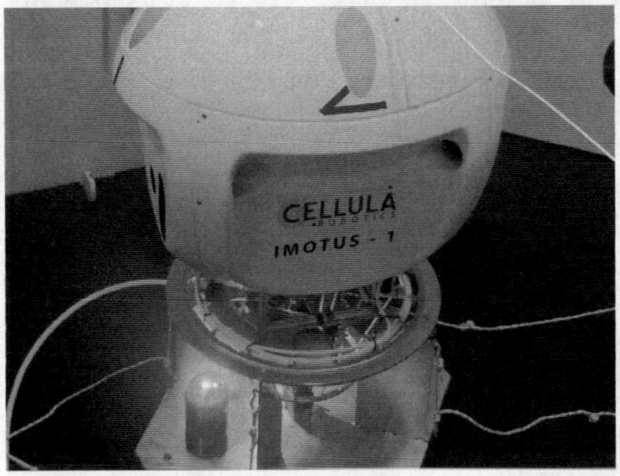

Acoustic and Optical Comms supporting AUV docking

OD OBN: Subsea data transfer with Saipem Flatfish

Sonardyne

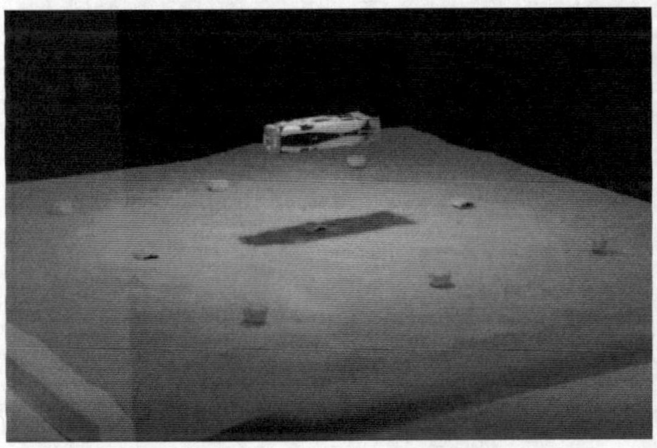

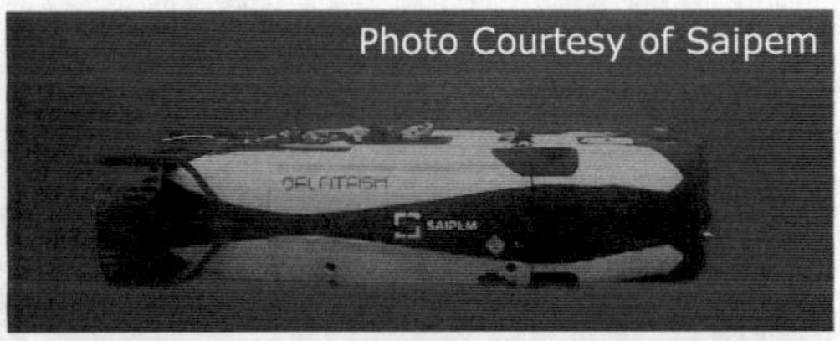

Acoustic location of On-demand Ocean Bottom Nodes (OD OBN)

Optical communication for high bandwidth data harvesting from OD OBN

And Finally... Smart Sound Connect Subsurface

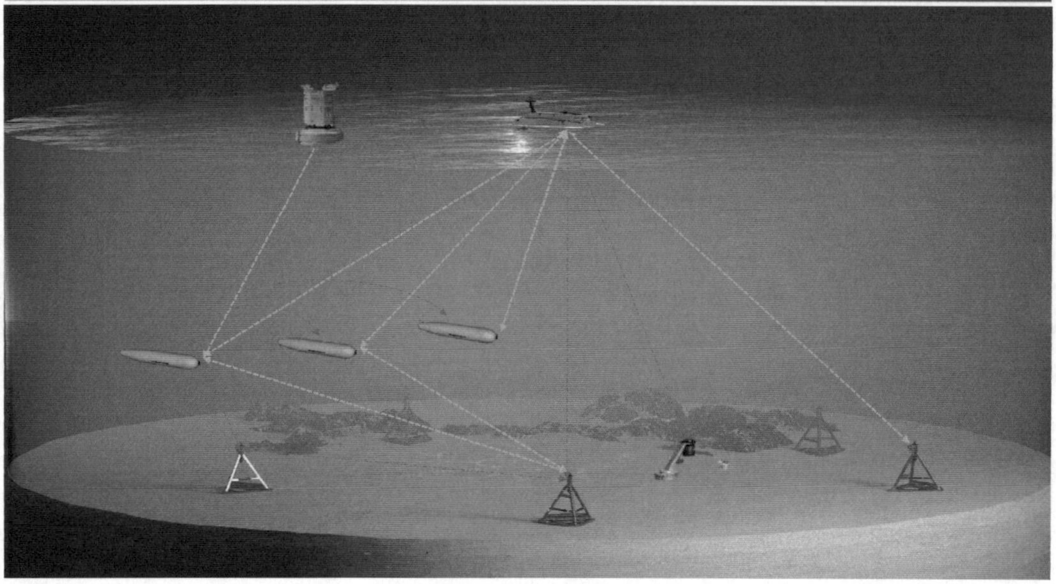

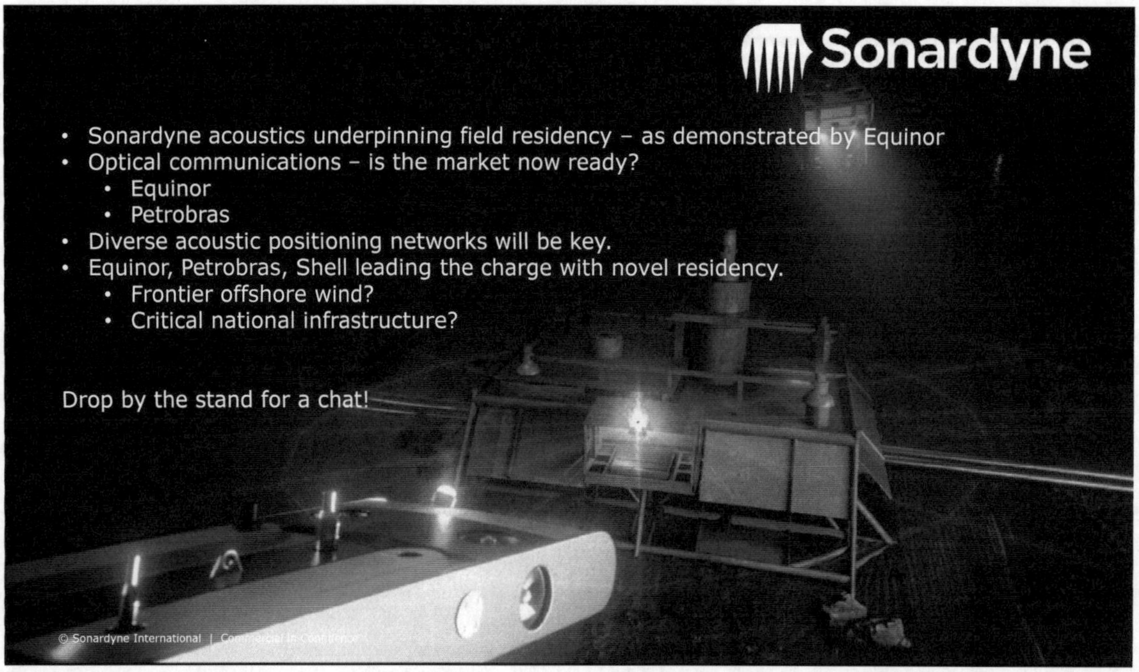

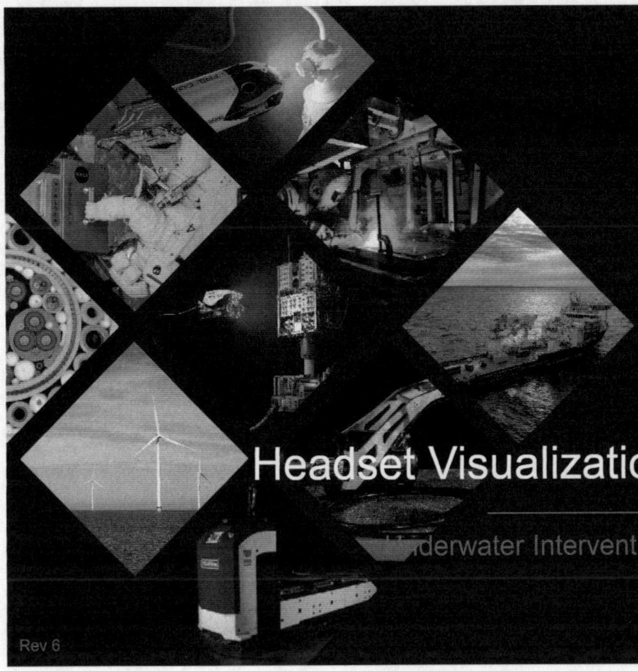

Headset Visualization for Work Class ROV

Underwater Intervention Forum – ROV Track Nov 12, 2024

Rev 6

Agenda

- Introduction to Oceaneering
- Next Generation ROV Service and the Need for Pilot Assistance
- Equipping a Work Class ROV to Leverage Mixed Reality
- VideoRay's OctoView for Mixed Reality Visualization
- Integrating Mixed Reality on a Work Class ROV
- Results of Initial Testing
- Future Steps in Developing Mixed Reality for Work Class ROV Control

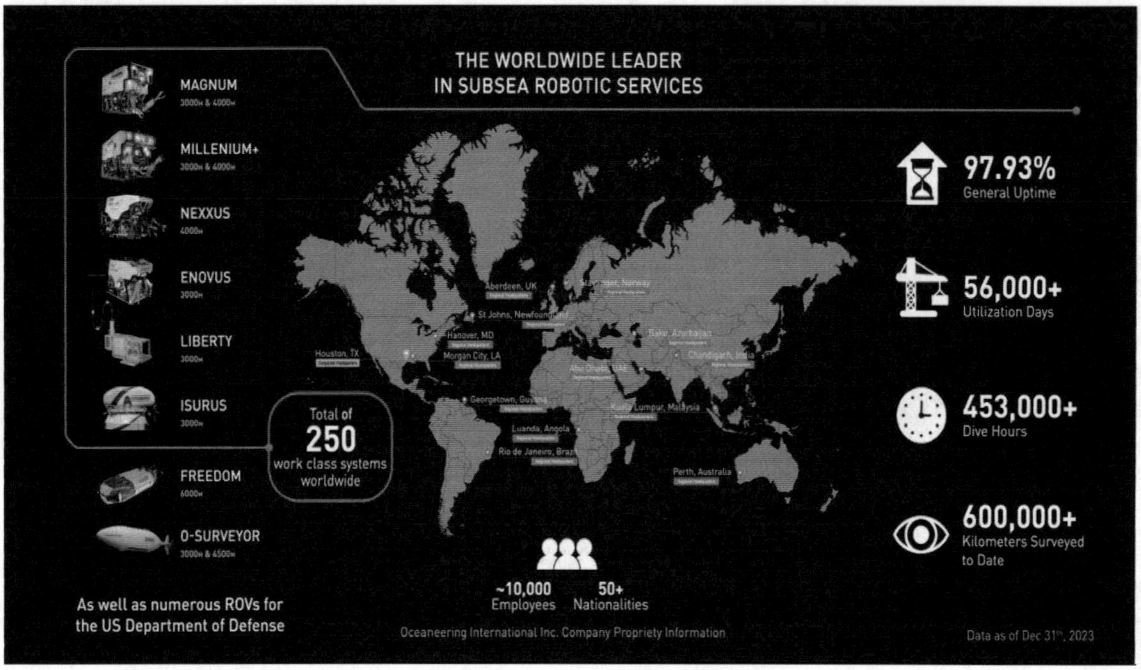

Next Generation ROV *SERVICE*

Why do we need to improve pilot skill?

According to (Kimberlite 2023 ROV Services Report)

- Pilot /Field Personnel Ranked significantly higher than ROV Quality/Reliability
- Pace of Operations are driven by Pilot Skill

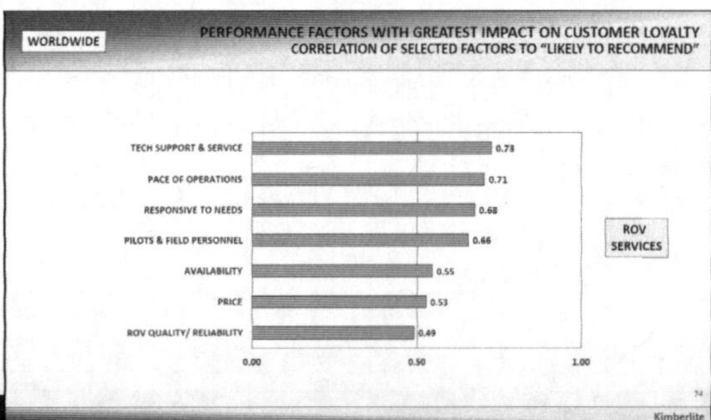

Damage to Equipment

- ~16% of Fleet Downtime relate to Tethers (2019-today)
- Tether failures typically result from pilot awareness issues

Consistency in Subsea Task Times

VideoRay OCEANEERING

The Need for Pilot Assistance

Equipping a Work Class ROV to Leverage Mixed Reality

Cameras

- 360 Cameras
  - Rear
  - Port Aft and Fore
  - Starboard Aft and Fore
- Forward Looking
  - Fixed Stereo Cameras
  - 4K Pan and Tilt
- Rear Facing
  - Tether Camera

Other Sensors

- Acoustic bumpers
- Scanning SONAR
- Altimeter
- DVL
- INS

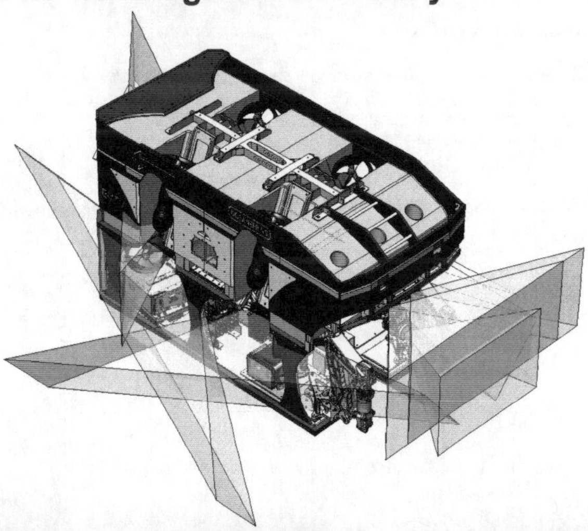

Existing VideoRay Projects Demonstrating Headset Visualization

Navy EOD PMS 408 and working with them on the MESR program of record

Monterrey Bay Aquarium Institute (MBARI)

Office of Naval Research Code 34 to utilize AR/VR for unmanned vehicle control of swarms through an SBIR Phase II.

VideoRay at a Glance

Company & Financial Overview

- Founded in 1999 with 50% growth YoY each of the last 3 years
- Recognized as the technological leader in "man portable" Remotely Operated Vehicles ("ROVs")
- Technical, engineering and customer support capabilities have positioned VideoRay with:
  - An exceptional reputation and brand heritage with over 20 years of industry leadership and innovation
  - 4,000+ ROVs shipped to-date across 75+ countries; one of the largest installed bases globally
  - A unique and diversified customer base includes the U.S. DoD, allied navies and Fortune 500 companies, among others
- Systems are uniquely suited for mission-critical applications in a variety of end-markets
- Headquartered in Pottstown, PA and has five facilities staffed

Introducing VideoRay's Leadership Team

OCEANEERING® VideoRay

Installed base of 4,000+ systems and growing global sales and support channels reinforce VideoRay's leadership position

VideoRay's Global Presence

4,000+
Total Systems Sold

1,600+
Customers

75+
Countries

30+
International Dealers

13
Authorized Servicing Centers

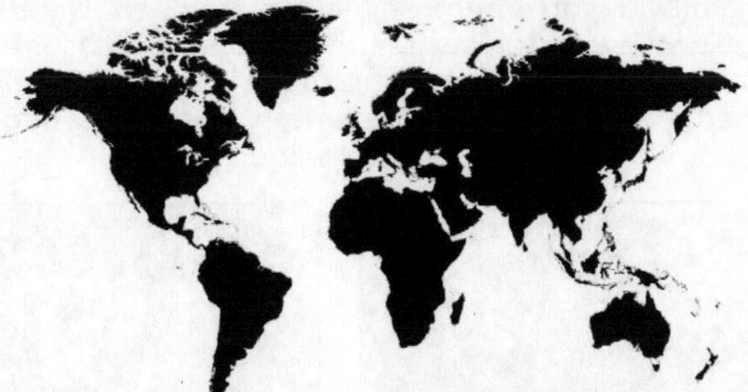

70%
Americas

24%
EMEA

6%
Asia Pacific

Notable Customers

ASI Marine
Chevron
ExxonMobil
POWER
OCEANEERING

Mixed Reality & the Future of Work

OCEANEERING® VideoRay

OctoView

Blue Ring provides an enterprise teleoperations software application called **OctoView**. OctoView is a **revolutionary HUD** for ROV, USV, UGV, and UAS operators for **visualizing** and **manipulating** unmanned vehicle sensor data, telemetry, and spatial information in Mixed Reality.

Illustration of OctoView to control ROVs

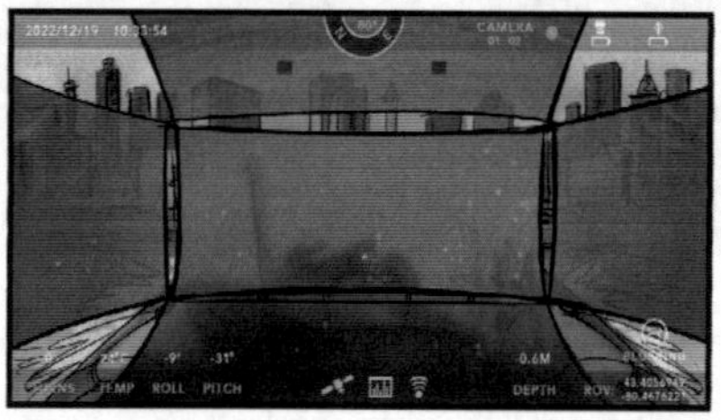

Illustration of an immersive interface which displays sensor information in spherical quadrants

OctoView

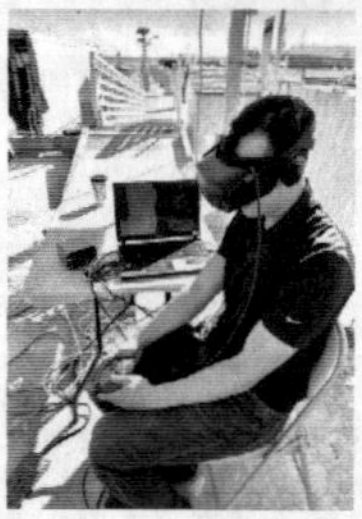

Integrating Mixed Reality into Work Class ROV Service

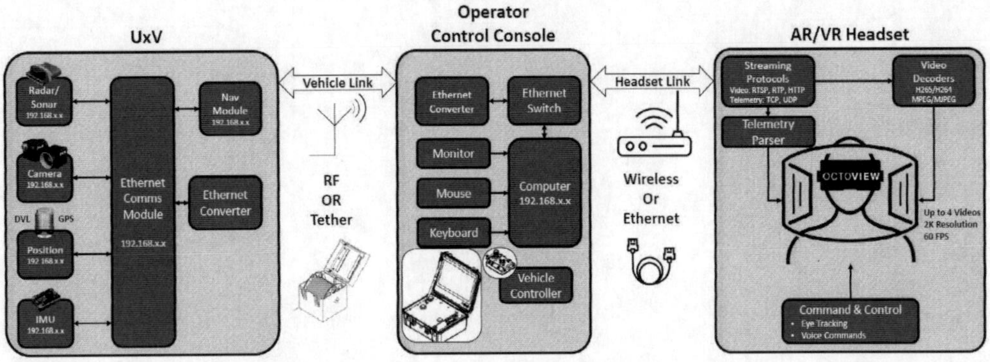

VideoRay OCEANEERING

Future Steps

- Stereo-Vision
- Augmented Reality
  - Enable Pilot to see Physical Controls
  - Control Integration Inside AR
- Operator Feedback
  - UFRJ-Macae Neurophysical Feedback Project

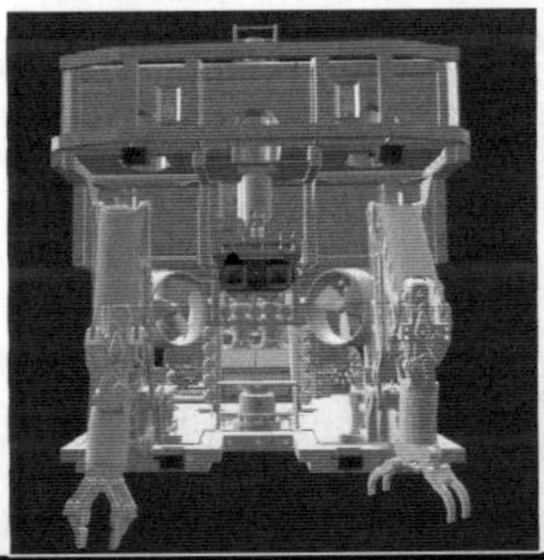

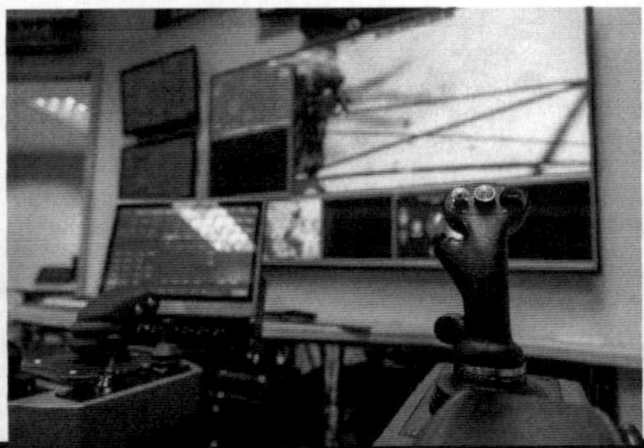

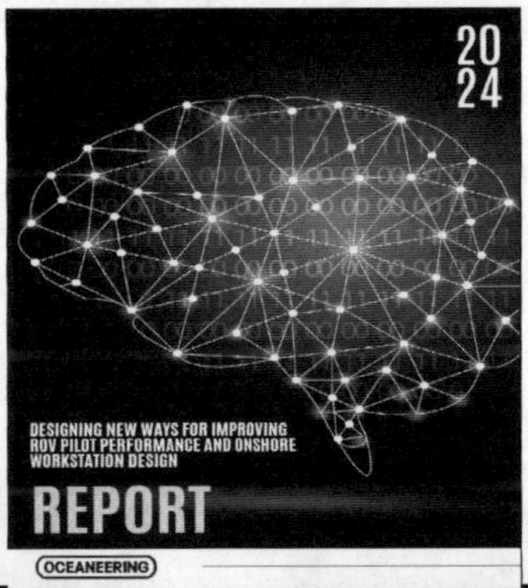

VideoRay OCEANEERING

Questions

Casey Sapp

VideoRay

VP of Strategy and Emerging Technology

casey.sapp@videoray.com

+1.407.718.9156

Nicholas Rouge

Oceaneering

Product Manager – Work Class ROVs

nrouge@oceaneering.com

+1.281.389.7284

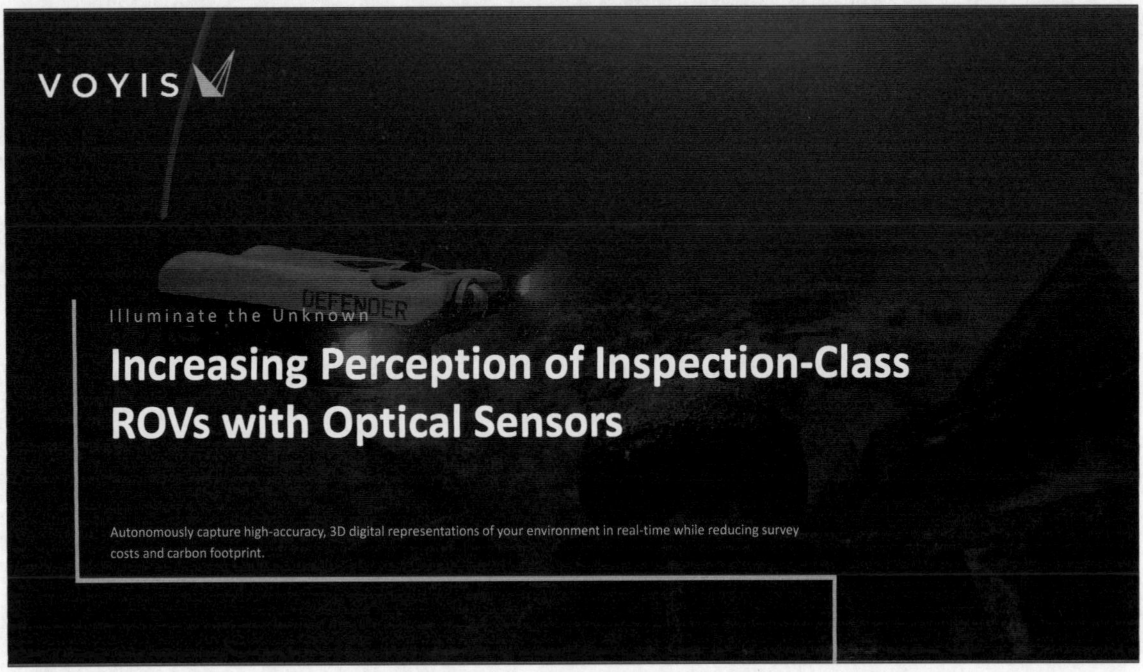

WELCOME

Agenda

1. About Voyis
2. Inspection Challenges
3. Voyis Optical Sensor
4. 3D Modelling
5. Case Studies

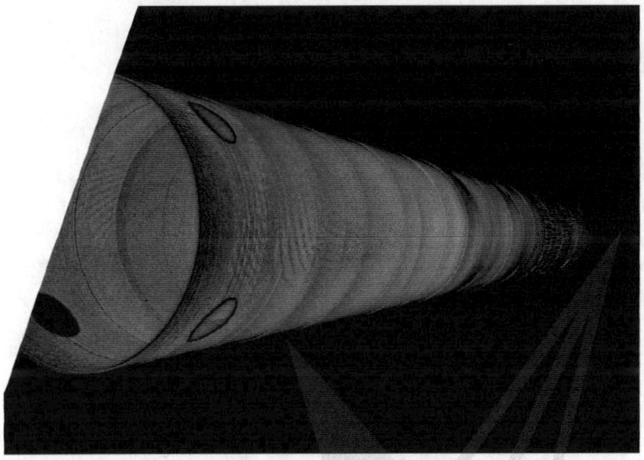

Russell Giles

4+ years of experience working in various Civil Industries including mining, deep foundations, construction & consulting

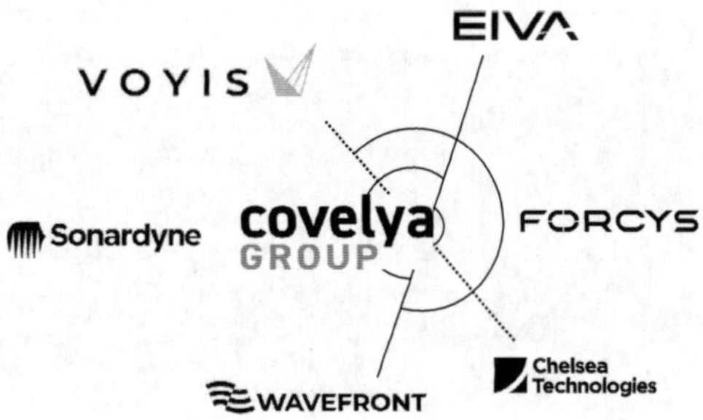

OUR TEAM

More Than Just Hardware

Our team is made up of passionate problem solvers who partner with our customers to consistently provide the best data possible.

- HQ in Waterloo, Ontario
- 1st Dynamic Laser Scanner for Subsea Vehicles (2011)
- Part of Covelya Group
- Laser & Camera package on Ernest Shackleton's Endurance

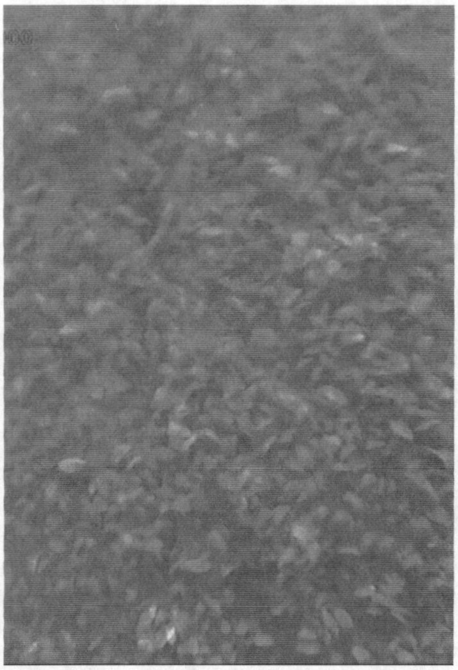

Inspection Challenges

Water Conditions

- Turbidity
- Current

Data Collection

- Coverage gaps
- Latency
- Repeatability

Data Processing

- Poor Data
- Confidence
- Time consuming

Project Costs

- Shutdowns
- Processing Time
- Client Satisfaction

OVERVIEW

Our Products

Standalone

Underwater Laser Scanners
Our high-resolution Insight laser scanners can be used for dynamic or static surveys and inspections.

Imaging Systems
Observer cameras paired with our Nova LED panels for crisp, evenly illuminated stills images.

Solutions

Recon AUV Payloads
A complete, power-efficient payload to expand your platform's capability.

Perception ROV Skids
Configurable inspection-class or WROV skids that suit your needs.

Discovery Systems

Imaging Systems

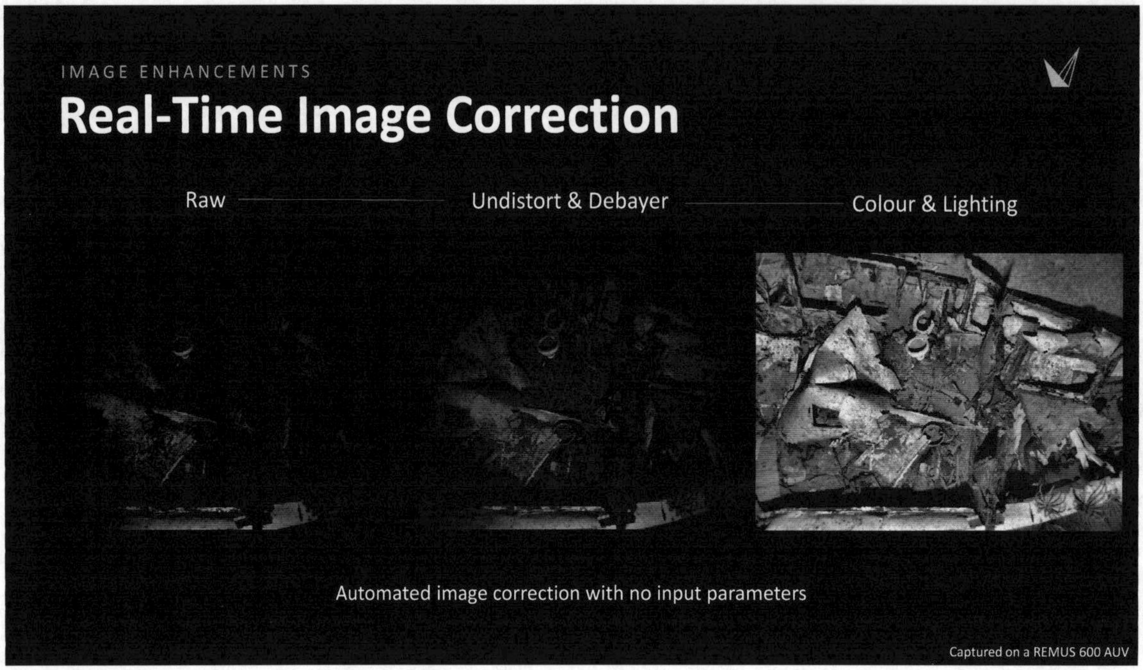
IMAGE ENHANCEMENTS
Real-Time Image Correction
Raw
Undistort & Debayer
Colour & Lighting
Automated image correction with no input parameters
Captured on a REMUS 600 AUV

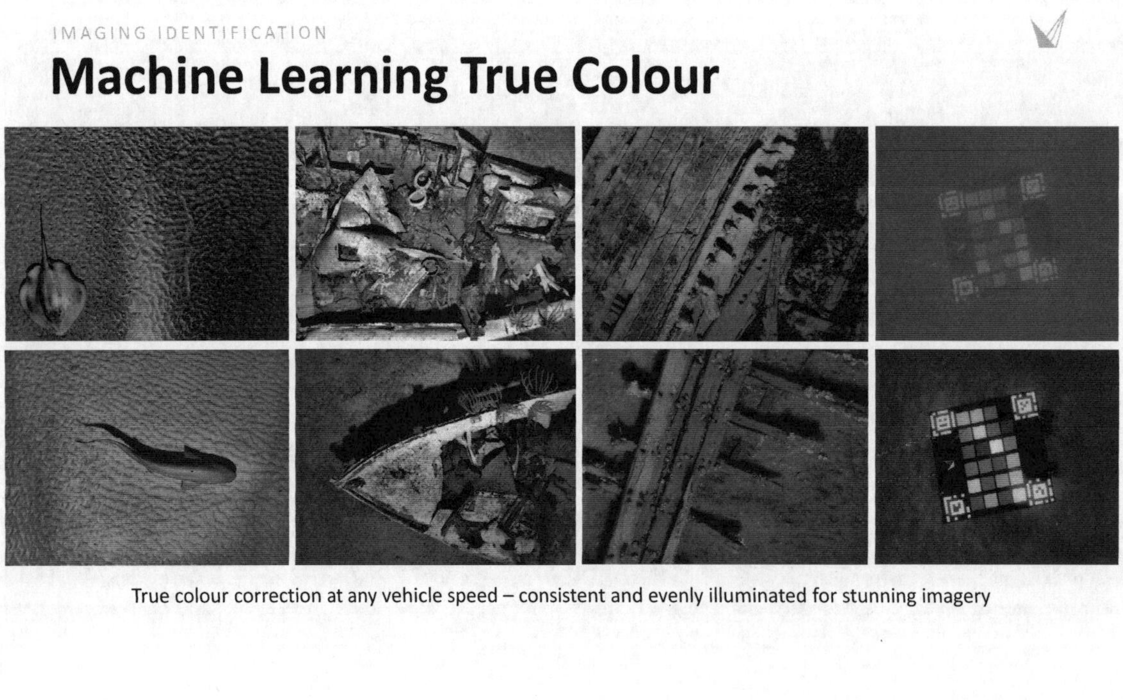
IMAGING IDENTIFICATION
Machine Learning True Colour
True colour correction at any vehicle speed – consistent and evenly illuminated for stunning imagery

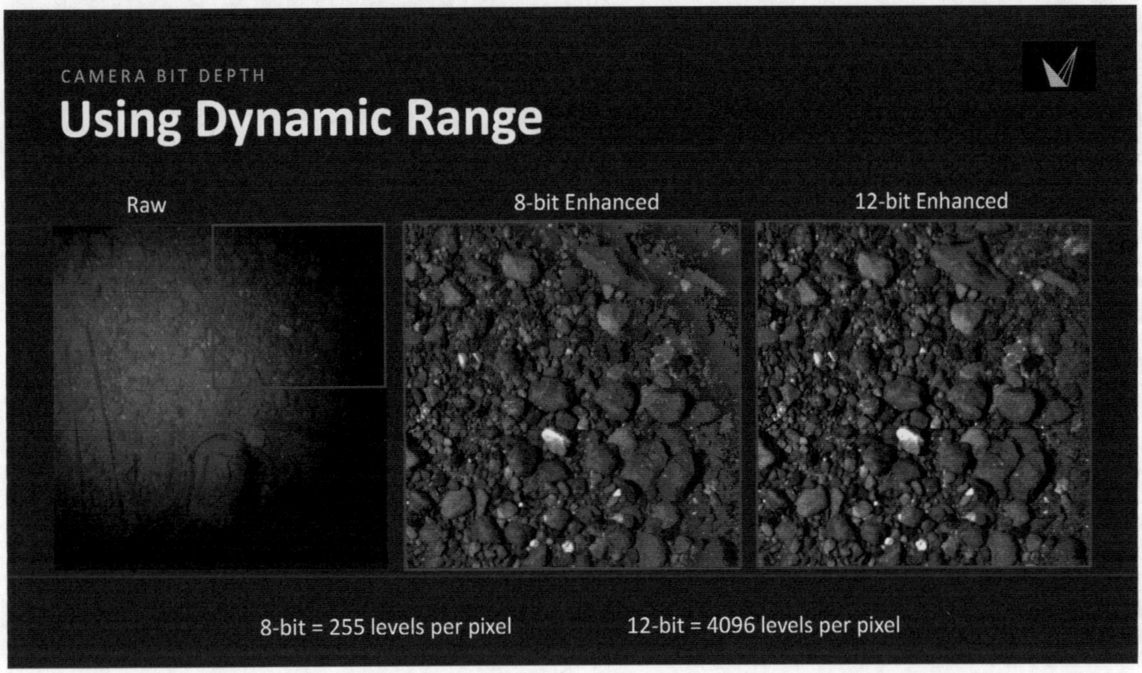
CAMERA BIT DEPTH
Using Dynamic Range
Raw
8-bit Enhanced
12-bit Enhanced
8-bit = 255 levels per pixel
12-bit = 4096 levels per pixel

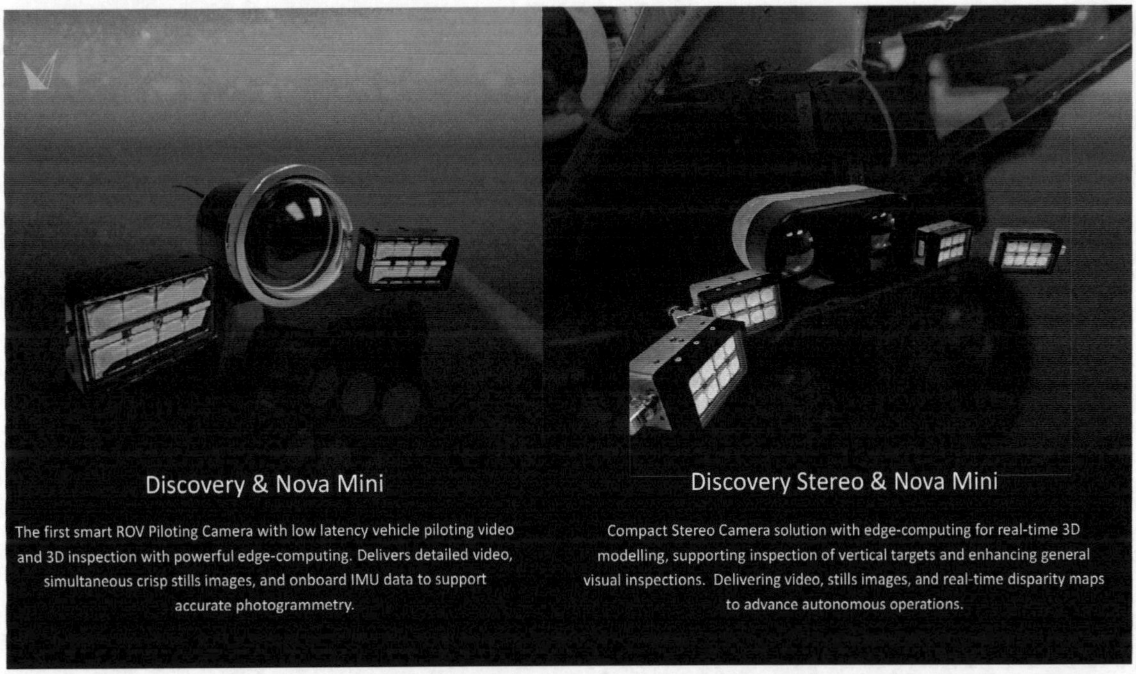
Discovery & Nova Mini
The first smart ROV Piloting Camera with low latency vehicle piloting video and 3D inspection with powerful edge-computing. Delivers detailed video, simultaneous crisp stills images, and onboard IMU data to support accurate photogrammetry.
Discovery Stereo & Nova Mini
Compact Stereo Camera solution with edge-computing for real-time 3D modelling, supporting inspection of vertical targets and enhancing general visual inspections. Delivering video, stills images, and real-time disparity maps to advance autonomous operations.

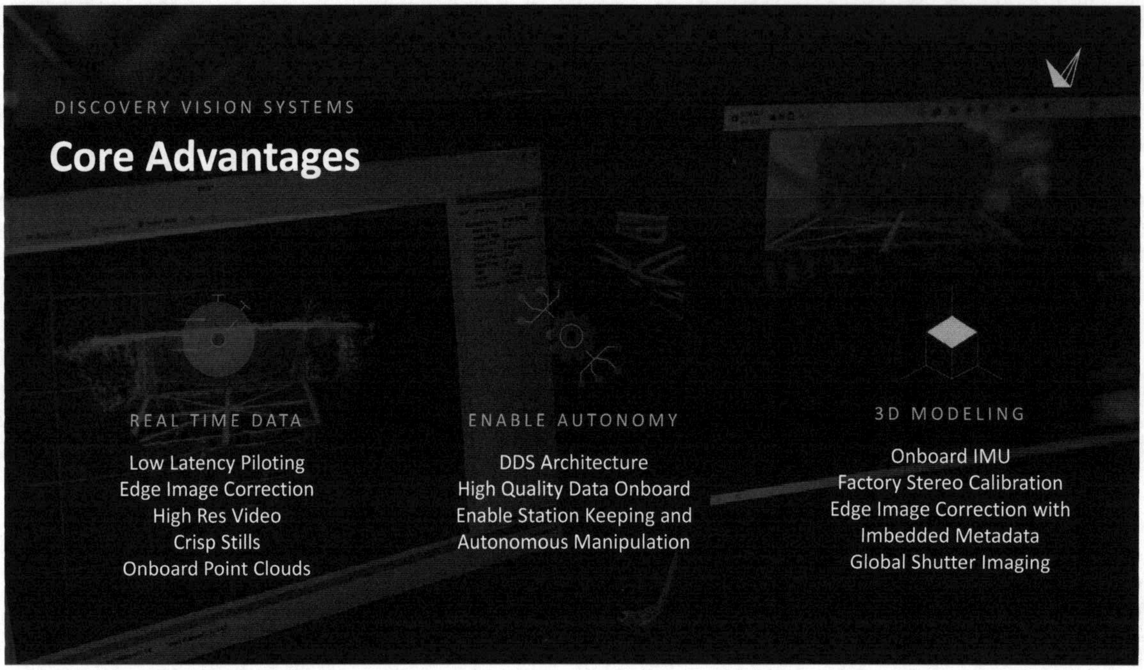
DISCOVERY VISION SYSTEMS
Core Advantages
REAL TIME DATA
Low Latency Piloting
Edge Image Correction
High Res Video
Crisp Stills
Onboard Point Clouds
ENABLE AUTONOMY
DDS Architecture
High Quality Data Onboard
Enable Station Keeping and
Autonomous Manipulation
3D MODELING
Onboard IMU
Factory Stereo Calibration
Edge Image Correction with
Imbedded Metadata
Global Shutter Imaging

Stills
Discovery Stills for Inspection Purposes

Pos: 0.0000000 0.0000000
Dpt: 5.4m
Alt: 0.0m

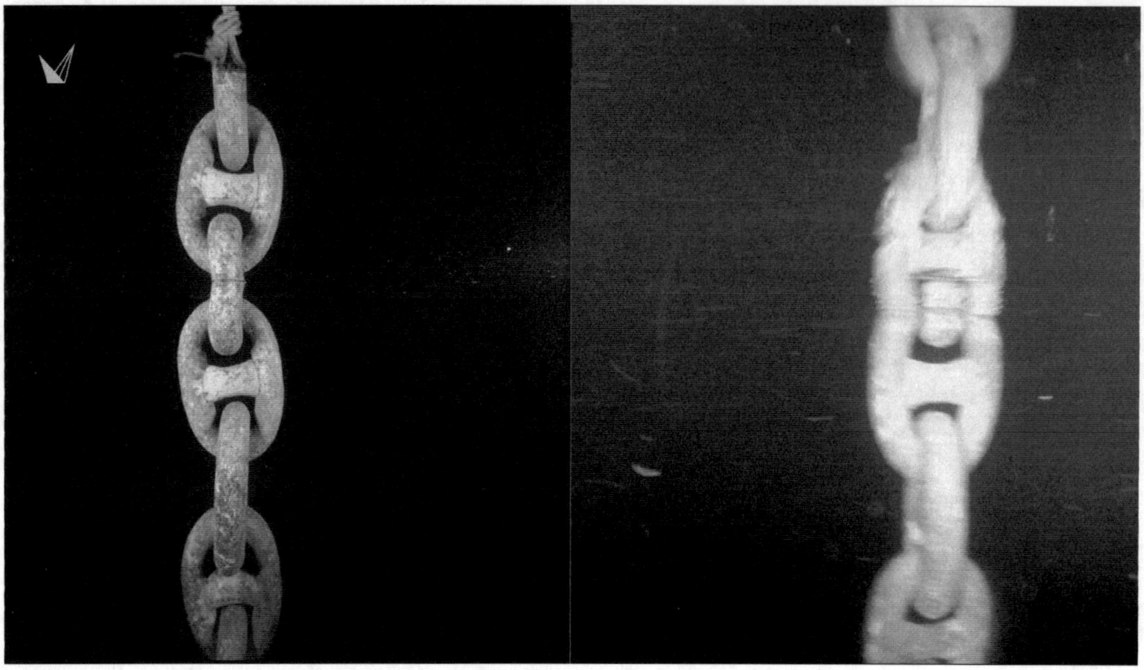

Case Studies

CASE STUDY

Ship Hull Inspection

Location:	Owen Sound, ON, Canada
Depth:	20 m
Vehicle:	Marinenav Ultimate
Equipment:	Discovery Stereo Camera
Objectives:	Generate 3D models of Ship Hull

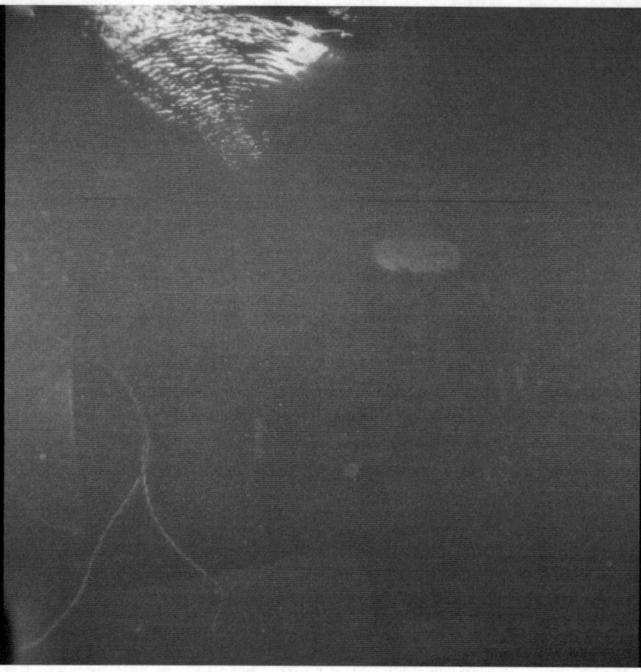

CASE STUDY

Pier Nose Inspection

Location:	NA
Depth:	10 m
Vehicle:	VideroRay Defender ROV
Equipment:	Discovery Mono Camera
Objectives:	Generate 3D models of ice damage to nose piers

21

CASE STUDY

OPG– Auburn Facility

Location:	Auburn Hydro Electric dam
Depth:	8m
Vehicle:	VideoRay Defender ROV
Equipment:	Discovery Mono Camera & Perception Skid
Objectives:	Complete 360 scans of trash racks, stop logs, concrete walls for a condition assessment

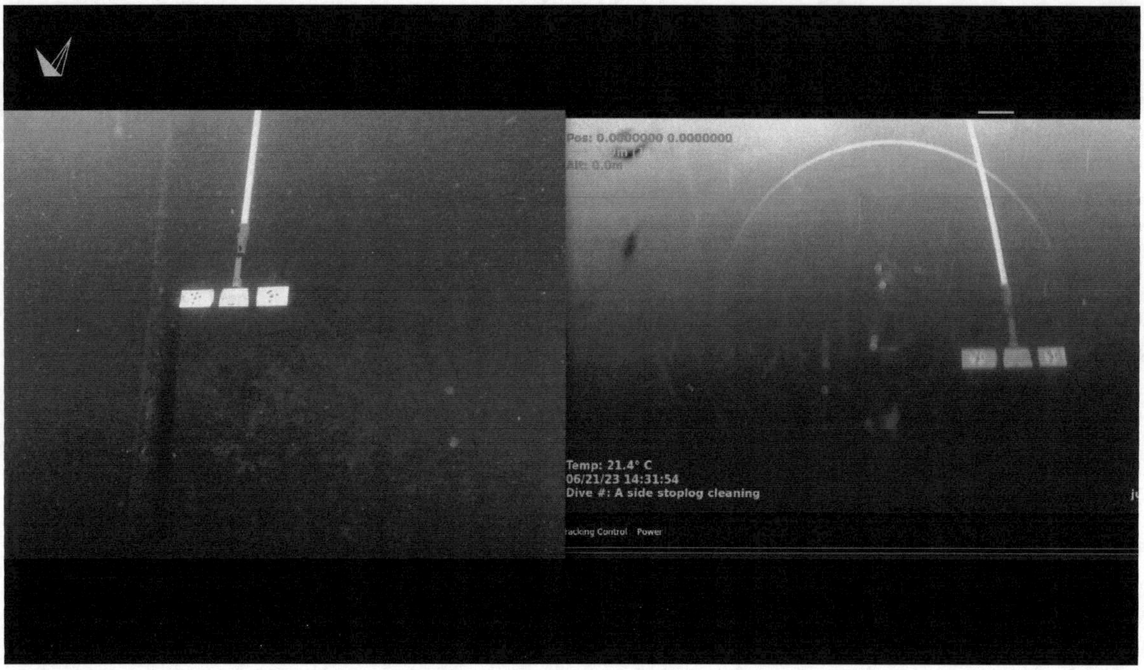
Pos: 0.0000000 0.0000000
Temp: 21.4° C
06/21/23 14:31:54
Dive #: A side stoplog cleaning

Sonar for Underwater Bridge Inspection
& How Changes in NBIS Affects Its Use

Brian Dilworth, P.E. – Collins Engineers, Inc.

Underwater Intervention - November 2024

UNDERWATER INTERVENTION
Powered by WORKBOAT

COLLINS ENGINEERS INC

Company Overview

- Established in 1979
- ENR Top 500 Design Firm
- Staff numbering 300+
- Licensed PEs and SEs
- LEED AP certified engineers
- In-house engineer-divers

COLLINS ENGINEERS INC

Continually on the leading edge of standards and methodology, Collins provides engineering ingenuity and solutions to complex problems, and realistic, honest answers.

My Background

- Purdue - Civil Engineer (Drew Breeeees!)
- 18 years of UW inspection experience
- ADCI Diving Supervisor
- Licensed PE in 10 states
- Trained in various UW imaging systems
- NHI Team Leader
- FHWA Instructor

COLLINS ENGINEERS INC.

Fun Fact: I performed an underwater facility inspection on the most-recent date of measurable snowfall in the City of New Orleans (Dec. 11, 2008)

Presentation Objectives

- To educate on sonar (acoustic imaging) for use in the bridge inspection industry
  - Who's in the audience... Inspectors? Infrastructure owners? Manufacturers? Sales?
- To inform on updates to NBIS related to underwater inspection & sonar
- Share my experience with sonar for UW bridge inspections

COLLINS ENGINEERS INC.

Presentation Objectives

- Spur conversation on experiences and interpretations
- To entertain
- To get you out early

COLLINS ENGINEERS

Disclaimer

- This presentation is intended to provide information to the audience regarding existing bridge inspection requirements that are current as of the time of presentation (Nov 2024).
- The contents of this presentation do not have the force and effect of law or regulation.
- Collins uses a variety of sonar systems from a variety of manufactures – there is no intent to promote one system over the other.

COLLINS ENGINEERS

Disclaimer

- The information and ideas presented are sourced from one of the following:
  - Collins Engineers, Inc.
  - Published, publicly-available information.
  - My own interpretation and opinions.
- Any opinions and (mis)interpretations of the Laws, Standards, Guidelines, etc. made in this presentation are that of Brian Dilworth, PE, ADCI, America's top-ranked NBIS UW Bridge Inspector

COLLINS ENGINEERS

NBIS Bridge Inspection Requirements

National Bridge Inspection Standards (NBIS)

- The Law since 1971 – found in Code of Federal Regulations (CFR)
- Sets **minimum** standards for proper safety inspection and evaluation of all highway bridges
- Significant number of updates in 2022
  - Most went into effect June 2022
  - Some sections were given 24 months to be in effect

Schoharie Creek Bridge Collapse (April 1987)

COLLINS ENGINEERS

Updates Related to UW Bridge Inspection

NBIS Definitions:

Team leader. *The on-site, nationally certified bridge inspector in charge of an inspection team and responsible for planning, preparing, performing, and reporting on bridge field inspections.*

Underwater bridge inspection diver. *The individual performing the inspection of the underwater portion of the bridge.*

COLLINS ENGINEERS

Updates Related to UW Bridge Inspection

NBIS Definitions:

Underwater inspection. *Inspection of the underwater portion of a bridge substructure and the surrounding channel, which cannot be inspected visually at low water or by wading or probing, and generally requiring diving or other appropriate techniques.*

COLLINS ENGINEERS

Updates Related to UW Bridge Inspection

New "Specifications for National Bridge Inventory" (SNBI), 2022

- *Developed in coordination with NBIS, AASHTO MBE, FHWA's BIRM*
- *Instructions for entering inspection data into National Bridge Inventory (NBI)*

COLLINS ENGINEERS

Updates Related to UW Bridge Inspection

SNBI – Section 6.2 – Item B.IE.12 – Inspection Equipment

Code A07 = Snorkel

Code A08 = SCUBA

Code A09 = Surface Supplied Air

Code A10 = Remotely Operated Vehicle (ROV)

Code I13 = Underwater Imaging

Commentary: *"Use code I13 when underwater imaging technologies such as side scan sonar are used to* ***supplement*** *underwater inspections."*

COLLINS ENGINEERS

Updates Related to UW Bridge Inspection

Examples provided in SNBI:

An underwater inspection was performed with divers using a boat and surface supplied air. Before the dive, side-scan sonar was performed to capture underwater images.

- *Report A06|A09|I13 for the underwater inspection.*

COLLINS ENGINEERS

Updates Related to UW Bridge Inspection

Examples provided in SNBI:

A scour critical bridge experienced flood water elevations up to the web of the exterior girder. Per the scour POA, scour monitoring was immediately completed by a team leader. A remotely operated water vehicle was used that was equipped with underwater imaging technology.

- *Report A10|I13 for the scour monitoring inspection.*

COLLINS ENGINEERS

Updates Related to UW Bridge Inspection

Further muddying the waters...

The Federal Highway Administration (FHWA), through the National Bridge Inspection Standards (NBIS), and through guidance provided in Technical Advisories, specifies that underwater members be inspected to the extent necessary to determine with certainty that their condition has not compromised the structural safety of the bridge. To achieve that certainty, bridge owners may have to employ one or more underwater inspection techniques. These techniques may include visual and tactile inspections by wading, diving, remotely operated vehicles (ROVs), underwater cameras, underwater imaging, depth sounding equipment, material sampling equipment, and other specialized inspection equipment as needed to determine underwater structural and streambed conditions.

COLLINS ENGINEERS

Updates Related to UW Bridge Inspection

Further muddying the waters...

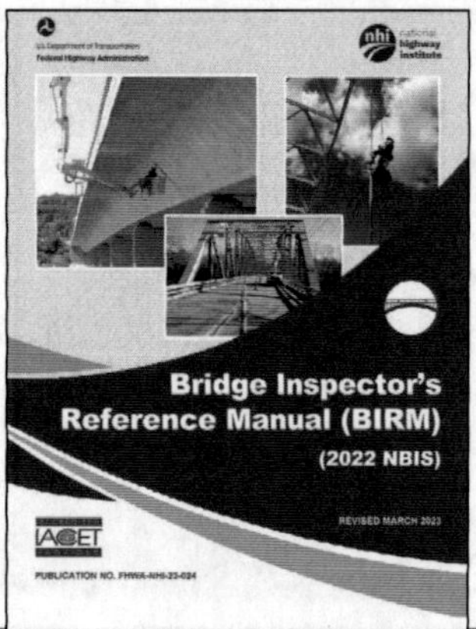

Underwater inspections necessitate trained underwater bridge inspection divers.

–FHWA/NHI Bridge Inspection Reference Manual, Chapter 2 (2023)

Routine Underwater inspections typically include a 100 percent Level I inspection and a 10 percent Level II inspection, but it may include a Level II and Level III inspection to determine the structural condition of any submerged portion of the substructure with certainty.

–FHWA/NHI Bridge Inspection Reference Manual, Chapter 16 (2023)

COLLINS ENGINEERS

Updates Related to UW Bridge Inspection

Further muddying the waters…

Consider the use of acoustical imaging technology in situations where water depth, high flow, turbidity, or other conditions could affect diver ability to determine localized conditions and scour effects. Use of this technology in the presence of submerged debris and other hazards provides safety benefits in lieu of deployment of a diver for Level I, II, or III inspections. Acoustical imaging technology could be a particularly effective technique for newer structures or where previous underwater inspections (Level I, II, or III) indicate no substructure element deficiencies. An appropriate mix of inspection techniques involving diving and nondestructive testing may be appropriate. States should develop policies and procedures for underwater inspection techniques; document these procedures; and include a schedule for Level I, II, and III inspections –AASHTO Manual for Bridge Evaluation (2022)

COLLINS ENGINEERS

Updates Related to UW Bridge Inspection

But alas… an answer to the question?

Underwater Q313-11 Can underwater imaging technology be used to supplement an underwater inspection? Can this technology be used instead of utilizing an underwater bridge inspection diver? (Updated 4/05/2024)

A313-11 Underwater imaging technology can supplement underwater inspections (UWI). These technologies provide owners a method of supplementing UWI by providing imaging of underwater elements, especially in cases where there are elevated safety risks in performing the UWI utilizing divers. As the imaging technology has advanced, FHWA recognized the need for a comprehensive evaluation of its capabilities, and in 2018 published the report "Underwater Inspection of Bridge Substructures Using Imaging Technology." The findings and conclusions of this report recommend the expanded use of sonar technology to improve the safety of bridges, including, but not necessarily limited to, Level I inspections, for broad characterization of bed conditions, and in conditions adverse to diving. However, it also concluded that sonar inspections have not demonstrated the ability to identify some smaller scale elements of substructure condition that may be important in assessing the bridge and recommending maintenance.

Based on this report, FHWA finds it acceptable to use imaging technology for the Level I portion of the UWI. The Level II portion of the UWI is still to be performed by an underwater bridge inspection diver. The combination of underwater imaging for Level I and diver for Level II will minimize the exposure time of the dive team. Attempts should be made to schedule the dive portion of the inspection during times when the risks are minimal. If there are diver safety issues which prevent performing the Level II portions of the UWI during the normally scheduled inspection, risks and consequences should be considered and a follow-up Level II inspection should be completed when it becomes safe to dive.

– FHWA/NHI website, Questions and Answers on the National Bridge Inspection Standards, 23 CFR Part 650, Subpart C

COLLINS ENGINEERS

Updates Related to UW Bridge Inspection

Why didn't you just say that on Slide 1 and be done?

- Diving (hands-on) inspection seems to be preferred method of NBIS "Underwater" Inspections
  - But sonar has its place
- NBIS (law) implies that if sonar is used, the person "inspecting" needs to be qualified to inspect structures underwater
- Team leader ultimately decides if the methods for inspecting are sufficient to determine structural condition below water
- Wanted a trip to New Orleans

COLLINS ENGINEERS

My Experiences using Sonar for Bridges

Benefit #1 (by a wide margin): SAFETY

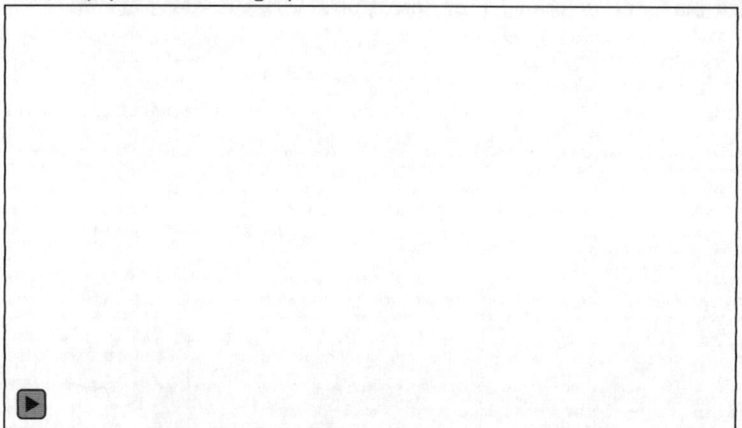

COLLINS ENGINEERS

My Experiences using Sonar for Bridges

Benefit #2: Increased Accuracy of UW Measurements

How accurate does the technology need to be for NBIS UW Bridge Inspections?

COLLINS ENGINEERS

My Experiences using Sonar for Bridges

Benefit #3: Debris or Riprap Survey

My Experiences using Sonar for Bridges

Benefit #4: Scour Inspection

Review of Presentation Objectives

- To educate on Sonar for use in the NBIS Bridge Inspection industry
- To inform on updates to NBIS
- Share my experience
- Spur conversation
- Entertain
- Get you out early

COLLINS ENGINEERS

Thank You

Brian Dilworth, PE
bdilworth@collinsengr.com

UNDERWATER INTERVENTION
Powered by: WORKBOAT

COLLINS ENGINEERS INC.

MTS
marine technology society
MATE ROV
COMPETITION
World
Championship
MATE FLOATS!
PROMOTING THE DEVELOPMENT OF NEW STUDENT
SKILLS THROUGH THE MATE ROV COMPETITION'S
MATE FLOATS! MISSION
CONTROL
BLADDER
PUMP
POWER
VALVES AND
SENSORS
BALLAST
Long Beach City College
mtsociety.org

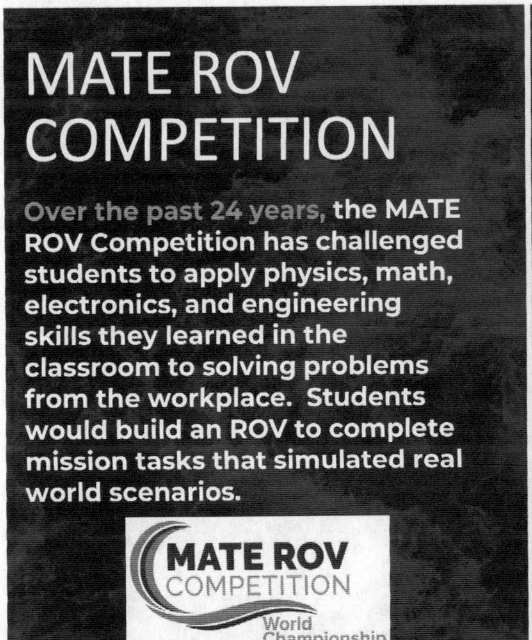
MATE ROV
COMPETITION
Over the past 24 years, the MATE ROV Competition has challenged students to apply physics, math, electronics, and engineering skills they learned in the classroom to solving problems from the workplace. Students would build an ROV to complete mission tasks that simulated real world scenarios.
MATE ROV
COMPETITION
World
Championship

MATE
FLOATS!
MATE FLOATS!
Over the past 4 years, the MATE ROV Competition, through the MATE Floats! mission, has challenged students to build an independent vertical profiling device, a float. This float moves down and up in the water column, collecting data, and when on the surface, transmitting that data back to a shoreside station.
Cabrillo Community College

Partner with GOBGC -

Global Ocean Biogeochemistry Array

MATE FLOATS!

The Global Ocean Biogeochemistry (GO-BGC) Array is a project funded by the US National Science Foundation to build a global network of chemical and biological sensors that will monitor ocean health. Scientists at the Monterey Bay Aquarium Research Institute, the University of Washington, Scripps Institution of Oceanography, Woods Hole Oceanographic Institution, and Princeton University plan to deploy 500 robotic ocean monitoring floats around the globe.

The MATE ROV Competition is part of the educational and outreach program of the Global Ocean Biogeochemistry Array.

GO-BGC

NSF

MATE Floats!

PROMOTING NEW STUDENT SKILLS

MATE FLOATS!

The goal of the *MATE Floats!* mission is to challenge students to build a vertical profiling float that encourages development of new skills for the blue economy and promotes knowledge about floats in general.

***MATE Floats!* started with just the "simple" task of building a profiler to go down and up in the water column. Each year, MATE has progressed to a new skill required from the float.**

What aspects promote the development of new skills?

- 2022: Vertical Profiles (down/up)
- 2023: Broadcast data to shoreside station
- 2024: Sensor data (depth/pressure)
- 2025: Depth sensor to hold position
- 2026 ????

Onboard Power

Onboard Power is not allowed for the ROV. But the float, operating independently, requires onboard power.

Skills and Knowledge needed:

- **POWERING THE PROFILER FROM BATTERIES (NiMH, Absorbed Glass Mat [AGM]) and battery limitations (max current allowed)**
- **SAFETY**
  - **Waterproofing**
  - **Mitigating water intrusion with overpressure relief devices**

MATE FLOATS!

Case Western Reserve University

Florida Atlantic University High School

Computer Control / Autonomy

MATE FLOATS!

The profiling float is completely independent of the ROV and shore side operations after launching. Once released into the water, the profiler is on its own and cannot be recovered to repair or reprogram.

Skills and Knowledge needed:

- **Programming**
- **Testing and Reliance – Need to be sure beforehand that the float will work reliably when deployed into the water. There are a number of tasks it must do, vertical profiles, gather sensor data, transmit to the shore.**

Active Buoyancy

Floats use **Active Buoyancy** to move vertically in the pool. Floats change their density by moving fluids in and out, which serves to change their volume.

Skills and Knowledge needed:

- **Physics - specific gravity, density, and how changes in those factors can move an object in the water column.**
- **Electronic systems – creating a system that controls a motor to move fluid into or out of the float (syringes, bladders, expansions)**

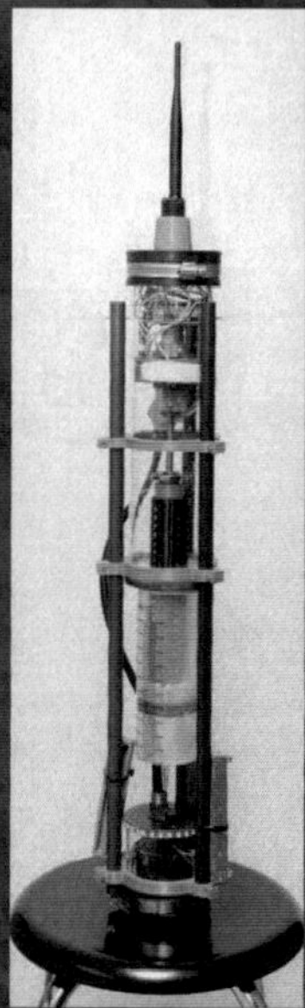

St. Francis Catholic High School

Purdue University

Getting data from the oceans to the shore, where scientists, researchers and the public can access and use it is a primary element of this decade's BLUE ECONOMY.

Sensor Data

In 2024, floats incorporated a **Depth/Pressure Sensor** and transmitted that data back to a shore side receiver. In 2025, the float will use that depth data to control position and hover **above the bottom of the pool.**

Skills and Knowledge needed:

- **Sensors – Incorporating sensors onto the independent float.**
- **Transmission – getting data from the float back to the mission station.**

21st Century Skills

Both the ROV and the Float promote **21st Century Skills.**

- Teamwork & Collaboration
- Problem Solving
- Critical Thinking
- Creativity
- Communication (written and oral)

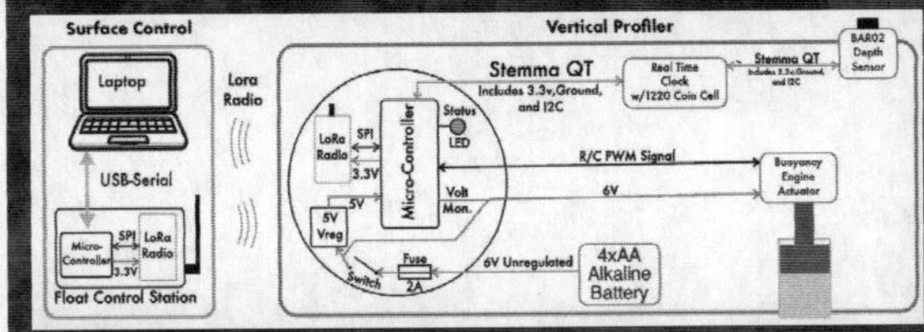

Estrella Mountain Community College – Float SID

MATE FLOATS!

Hong Kong University of Science & Technology

NAVIGATOR and SCOUT CLASSES

MATE FLOATS!

NAVIGATOR and SCOUT class are our younger participants, those just getting their feet wet in the world of underwater robotics. These tend to be upper elementary, middle and beginning high school students without any robotics backgrounds.

Early years focused on RECOVERING A FLOAT and GRAPHING DATA. The MATE ROV Competition also introduced them to ADOPT-A-FLOAT, the other aspect of GOBGCs education and outreach, where teams can adopt a float, name it, and provide drawings and designs to go on the outside of the float.

IN 2024, NAVIGATOR and SCOUT class students built a TETHERED PROFILER with a TEMPERATURE SENSOR. This profiler was manually driven down and up in the water column and would collect temperature at various points along the way.

IN 2025, teams are challenged to go autonomous. Program the vertical profiler to go from the surface, to the bottom, and back to the surface.

NAVIGATOR and SCOUT

IN 2025, teams are challenged to go autonomous. Students will program the vertical profiler to operate independently. It will still be a tethered profiler powered from the surface, but teams will program an Arduino, Raspberry Pi or other device to have the profiler move AUTONOMOUSLY down to the bottom and then return to the surface.

Skills and Knowledge needed:

- Programming autonomous operations
- Control systems

MATE FLOATS!

The SeaMATE Vertical Profiler

Conclusion

MATE FLOATS!

The inclusion of MATE Floats! into the MATE ROV Competition has allowed, and will continue to allow, students to learn new concepts and gain new skills that will help them in the 21st century blue economy. It offers them a chance to build on the technical skills they learned while building an ROV and tackle this new mission task with the creativity and ingenuity required for today's ocean technology workforce.

MTS
marine technology society

MATE FLOATS!

QUESTIONS?

Matthew Gardner

Marine Technology Society

MATE ROV Competition – Competition Technical Manager

mgardner@marinetech.org

mtsociety.org

MATE Floats! Marine Technology Summer Workshop

Hosted by the University of Washington School of Oceanography

INSTRUCTED BY: Rick Rupan, Sasha Seroy, Katy Christensen**, Paige McKay*, Tegan Yao*, and Riley Smith*

PRESENTED BY: Paige McKay

*Undergraduate student mentors, **Graduate student instructor*

Program Structure

- **Monday:** Introduction to oceanographic data, sensors, and vehicles including an ARGO Float Lab tour to introduce Floats
- **Tuesday:** Introduction to Python coding and a lab skills learning opportunity
- **Wednesday:** Sensor design and skills lab, deploying temperature sensors in Lake Washington and analyzing this data in Python
- **Thursday:** Half of the day aboard the R/V Rachel Carson, the other half learning career development skills
- **Friday:** Cruise Data analysis in Python and a resume workshop

Program overview and goals

- **Understand** how GO-BGC float technology functions to monitor changes in the global ocean.
- **Build sensors** to collect and assess oceanographic data
- **Utilize technology** to collect oceanographic and/or engineering data in support of scientific and/or engineering hypotheses.
- **Analyze** and i**nterpret** GO-BGC float data using Python programming.
- **Collaborate** to conduct studies, **communicate** data professionally, and **prepare for careers** in ocean technology

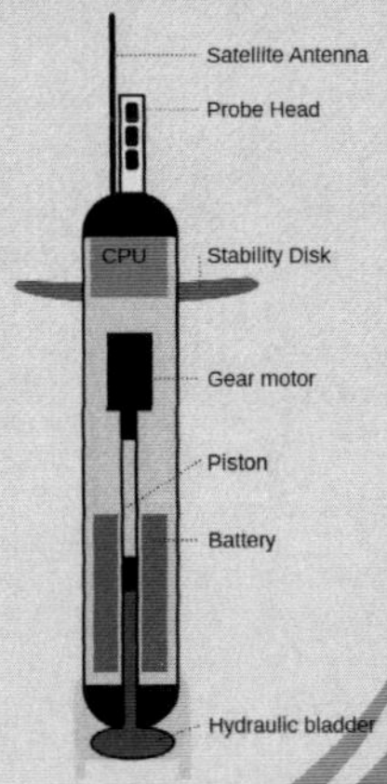

NCAT MTS marine technology society GO-BGC W UNIVERSITY of WASHINGTON

MATE Floats! Student Speakers

- Stephanie L'Heureux -
  - Cabrillo College
- Siena Hemminger -
  - San Diego Mesa College
  - UC San Diego
- Angela Nguyen -
  - San Diego Miramar College

NCAT MTS marine technology society GO-BGC W UNIVERSITY of WASHINGTON

MATE FLOATS
MARINE TECHNOLOGY WORKSHOP

Stephanie L'Heureux

CABRILLO COLLEGE

Computer Science && Ocean Engineering

NCAT MTS marine technology society GO-BGC UNIVERSITY of WASHINGTON

ABOUT ME

- Student at Cabrillo College, Aptos CA
- Studying Computer Science and Ocean Engineering
- Expected transfer Fall 2026
- MATE ROV Competition participant
- President of Cabrillo Robotics Club

NCAT MTS marine technology society GO-BGC UNIVERSITY of WASHINGTON

WHY MATE FLOATS?

My involvement in the MATE Floats program stemmed from a long-standing interest in marine technology and my participation in the MATE ROV Competition. I was a **MATE ROV competitor from 2018-2024**, an experience that shaped my career aspirations to become a **software developer for underwater vehicles**. I learned about the MATE Floats program through the MATE ROV and was intrigued by the opportunity to gain additional hands-on experience with different types of marine technology (**GO-BGC floats**) and **time aboard a research vessel**.

NCAT MTS marine technology society GO-BGC UNIVERSITY of WASHINGTON

SHIP EXPERIENCE

TIME ABOARD A RESEARCH SHIP

DATA COLLECTION INSTRUMENTATION

LAB SKILLS

DATA ANALYSIS

MATE FLOATS!
IMPACT
LOOKING AHEAD
During my time attending MATE Floats, I gained skills and experiences that will be directly applicable to my career including **lab skills** and time aboard a **research vessel**.
I also had the opportunity to build **meaningful connections** with like-minded individuals, many of whom I am still in touch with and who are here today.
I plan to graduate with a B.S. in **Computer Science** and an M.S. in **Ocean Engineering**. My career goal is to specialize in developing software for underwater vehicles.
NCAT
MTS
marine technology society
GO-BGC
W
UNIVERSITY of WASHINGTON

THANK YOU
INTERNATIONAL WORKBOAT SHOW
MTS
marine technology society
GO-BGC
NCAT
MATE FLOATS!
W
UNIVERSITY of WASHINGTON
Cabrillo College

About Me!

- From Riverside CA, living in San Diego
- Third Year Transfer from Mesa College at UC San Diego
- Intended Graduation Date: June 2026
- Studying Geosciences through the Scripps Institute of Oceanography at UC San Diego

On board the Rachel Carson research vessel

Tioga Pass in East Yosemite

NCAT MTS marine technology society GO-BGC W UNIVERSITY of WASHINGTON

Why the MATE Floats! Summer Workshop?

- Inspired by SDAG meeting where students presented research and were on JOIDES Resolution Expedition 379T
- Interest further sparked through learning about Scripps and UW oceanography departments
- Led to finding MATE Floats!

JODIES Resolution Vessel

2024 MATE Floats members in front of the Rachel Carson Research Vessel

Argo Float Workshop

NCAT MTS marine technology society GO-BGC W UNIVERSITY of WASHINGTON

Workshop Highlight - Chlorophyll Levels in the Water Column

- Recorded data from trip on Rachel Carson research vessel in Puget Sound
- Water column samples from Niskin Bottles -> Filtering -> Using a fluorometer to measure chlorophyll levels

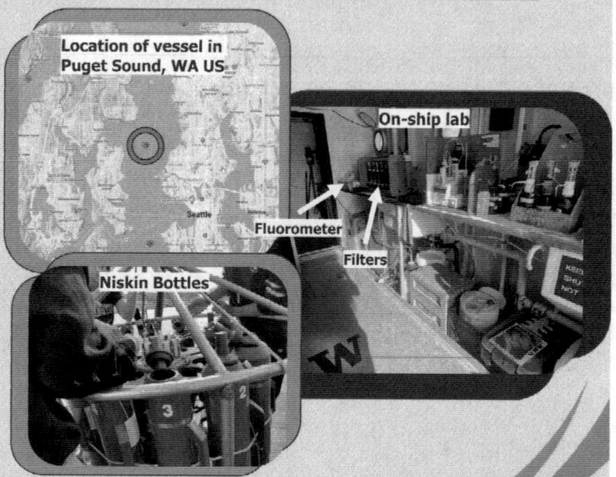

NCAT MTS marine technology society GO-BGC UNIVERSITY of WASHINGTON

Comparing Chlorophyll Levels from May to August

Question: Did Chlorophyll Levels change from May to August?

- Chlorophyll levels were higher in May than in August at the surface, similar at depth.
- Light levels were the same both months
- Suggests something other than light influences chlorophyll

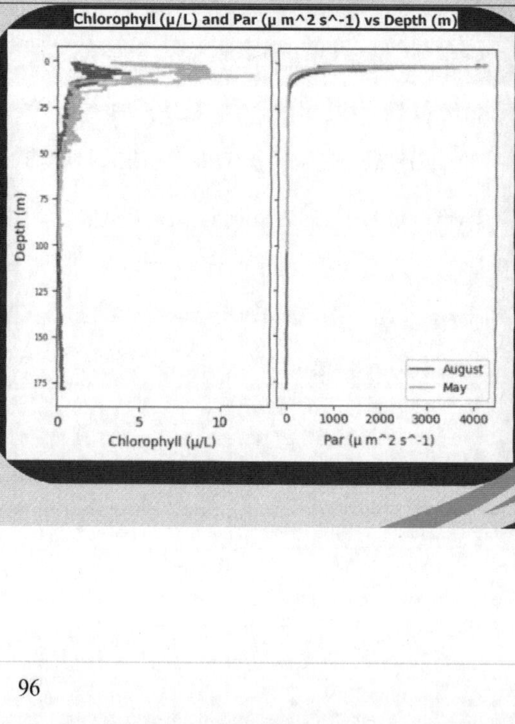

NCAT MTS marine technology society GO-BGC UNIVERSITY of WASHINGTON

Looking Ahead for Me

- This program inspired me to continue on an ocean and natural science path for my future career
- Also I am inspired to go into higher education
- Looking forward to hopefully joining a lab working in the Scripps Institute of Oceanography
- Ultimate goal is to have a career in Marine Geology, General Geoscience, Hydrogeology or something related

MATE FLOATS
MARINE TECHNOLOGY WORKSHOP

Thank you to all of my professors at the San Diego Community College District for teaching and inspiring me.

“ The sea is everythi
- Jules Verne, Twent
Thousand Leagues
Under the Sea

MATE FLOATS
MARINE TECHNOLOGY WORKSHOP

MATE Floats! Summer Workshop

PRESENTED BY:

Angela Nguyen

NCAT MTS marine technology society GO-BGC W UNIVERSITY of WASHINGTON

About me

- Student at San Diego Miramar College
- Intend to transfer as an Electrical Engineering major in Fall 25
- Hobbies: Running, hiking, anything outdoors!

NCAT MTS marine technology society GO-BGC W UNIVERSITY of WASHINGTON

Why MATE Floats?

- Scripps Institute of Oceanography Influence in SD
- Curious about research done in oceanography
- Career goal: Power Engineer, want to be interdisciplinary "trained"

NCAT MTS marine technology society GO-BGC W UNIVERSITY of WASHINGTON

Workshop Highlight: Cruise Data Analysis

Location of Cast

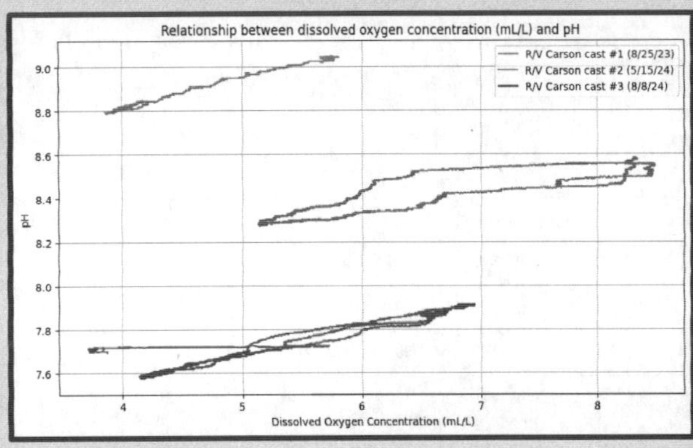

NCAT MTS marine technology society GO-BGC W UNIVERSITY of WASHINGTON

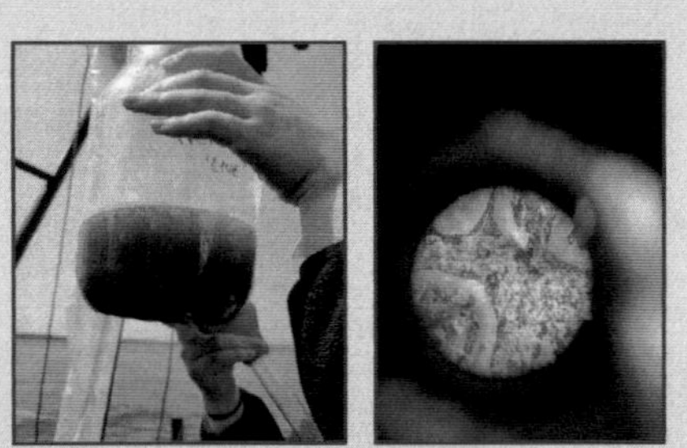

Puget Sound Water collection

ESP32 microcontroller

Chlorophyll collection, titrations, CTDs

Impact

- ←*Really* supportive and helpful students and faculty
- Realized how interdisciplinary EE is + will jump on any opportunity to go on another R/V!
- Learned a lot about data analysis in Python (Pandas, NumPy, MatPlotLib)

NCAT MTS marine technology society GO-BGC UNIVERSITY of WASHINGTON

RACHEL CARSON

Ocean Exploration Video Challenge

18 Approaches to Solving Organism Visual Tracking

Patrick Cooper & Eoghan Mcivor

November, 2024

OCEAN EXPLORATION

MATE ROV COMPETITION

What We're Covering

1. Introduction to the Challenge
2. Technical Approaches
3. Winning Designs
4. Takeaways

Ocean Exploration and Research

Introduction to the Challenge

Partnership: MATE ROV Competition and The NOAA Office of Ocean Exploration
Goal: Streamline the annotation process of organisms in ROV dive videos
Focus: Development of AI models to track and annotate a specific class of organisms (brittle stars) in video footage
Challenge Video: Continuous tracking shot of the ocean floor by an ROV

Ocean Exploration Video Challenge

The MATE ROV Competition is excited to partner with NOAA Ocean Exploration again this year to offer up this challenge to students from around the world!

NOAA Ocean Exploration Background

NOAA Ocean Exploration is dedicated to exploring the unknown ocean, unlocking its potential through scientific discovery, technological advancements, partnerships, and data delivery. We lead national efforts to fill gaps in our basic understanding of the marine environment, providing critical ocean data, information, and awareness needed to strengthen the economy, health, and security of the United States and the world. Through live video streams, online expedition coverage, training opportunities, and education and outreach programming, we enable scientists, resource managers, students, educators, members of the general public, and others to actively experience ocean exploration, broadening scientific participation, cultivating the next generation of ocean explorers, and engaging the public in exploration activities.

The data and information collected during expeditions and research supported by NOAA Ocean Exploration are publicly available, aiding stakeholders in identifying, understanding, and managing ocean resources for today and the future.

Ocean Exploration and Research

NOAA Ocean Exploration Background

Mission: Explore unknown ocean environments, unlocking scientific discovery and data for broad use
Activities: Live video streams, public outreach, and data delivery through expeditions
Objective: Aid in understanding and managing ocean resources with publicly available data. Experts and scientists annotate video in real-time or via recordings

Ocean Exploration and Research

The Data

The **Deep Discoverer** (D2) is a remotely operated vehicle (ROV) developed and maintained by the Global Foundation for Ocean Exploration (GFOE) for NOAA Ocean Exploration. It operates in tandem with its companion vehicle, Seirios, aboard the NOAA Ship Okeanos Explorer.

High-Definition Imaging: Equipped with a primary camera capable of zooming in on a 3-inch (7.6-centimeter) organism from a distance of 10 feet (3 meters).

Ocean Exploration and Research

Challenge Task and Data

Problem: Annotation of organisms is a time-consuming, manual process

Task:

- Build an AI model to track and annotate a specific organism (brittle stars) in video footage
- Maintain accurate bounding boxes while tracking throughout the video

Data:

- Continuous tracking shot from ROV footage (dynamic, ocean floor)
- **Training Data**: Available at FathomNet, with the decision to either focus on highly specific data or generalize the dataset.

Ocean Exploration and Research

Submission and Requirements

Submission Deadline: August 25, 2024, 11:59 PM (Hawaii time)

Documents Needed:

- Program with user documentation
- Resulting spreadsheet
- Video with bounding boxes

File Naming Convention: School/Organization_Name_Document_Type_2024

Testing: Model tested on video clips with similar data

Model Formats: .pt, .pth, TensorFlow 2.x, or .HDF5

Code Documentation: Must follow PEP 8 and be well-commented

MTS marine technology society

MATE ROV COMPETITION

Ocean Exploration and Research

Judging Criteria

Accuracy of Bounding Boxes:

1. Temporal Persistence
2. Bounding Box Accuracy
3. Amount of False Positives
4. Amount of False Negatives

Ocean Exploration and Research

Prizes

Winner Announcement: September 2024
Top 3 Teams: Posted on MATE ROV website
Prize: Travel stipend to the 2024 Underwater Intervention conference, New Orleans, LA, USA

Ocean Exploration and Research

2. Technical Approaches

Base Model Distribution

- 17 CNN-based Architectures vs 1 Transformer-based Architecture
- 13 Used Specific Training Data, 5 Performed General Unsupervised Pretraining
- Many Used Foundation Models to Generate Training Data
- Some Used Simulated Imagery to Pretrain Their Models

Ocean Exploration and Research

Base Models Deployed

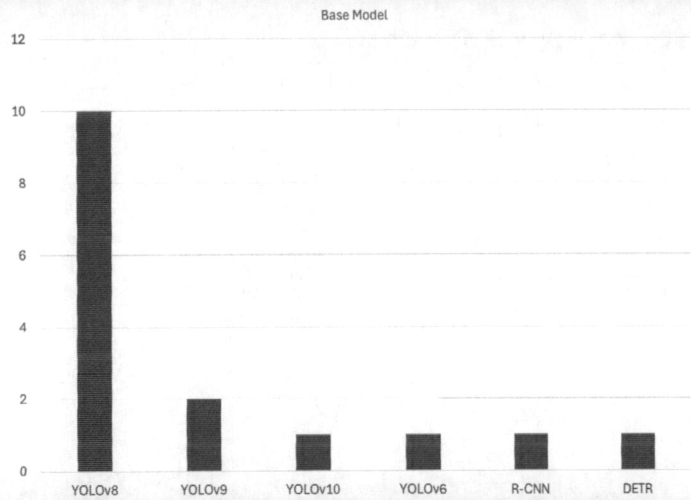

Ocean Exploration and Research

Introduction to YOLO (You Only Look Once)

Overview: YOLO is a real-time object detection architecture developed by Joseph Redmon in 2015.
Goal: Achieve high-speed and accurate object detection by processing images in a single pass.
Difference: Unlike traditional methods (e.g., R-CNN, Fast R-CNN) that use multi-stage pipelines, YOLO treats detection as a single regression problem.

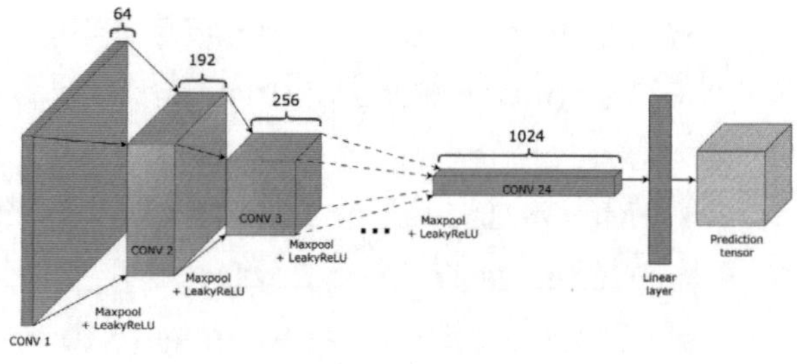

Ocean Exploration and Research

YOLO Architecture

Single Convolutional Network: Divides the input image into a grid and predicts bounding boxes, confidence scores, and class probabilities for each cell. Then YOLO uses IOU in non-maximum suppression to eliminate redundant or overlapping boxes, ensuring that only the bounding box with the highest IoU is kept for each detected object.

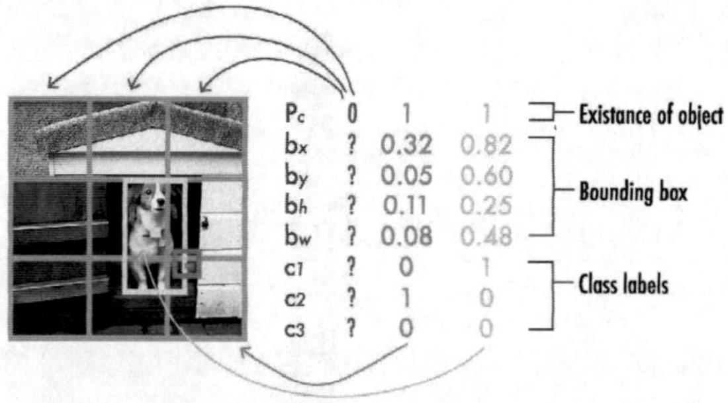

Ocean Exploration and Research

YOLO Architecture

- **Output - Bounding Box Coordinates**: (x, y, width, height)
- **Confidence Score**: Likelihood of object presence
- **Class Probabilities**: Multi-class predictions per grid cell
- **Efficiency**: Processes images in a single pass, achieving up to 45 frames per second (FPS).

Ocean Exploration and Research

Advantages and Evolution

Advantages

- **Speed**: Real-time performance, ideal for applications like autonomous driving and surveillance (and organism annotation).
- **Global Context**: Considers entire image context, reducing background errors.
- **Evolutions**: Subsequent models (YOLOv2, YOLOv3, etc.) have improved both accuracy and processing speed.

Ocean Exploration and Research

Model Comparison: E-JUST

Top-Performing Model: YOLOv8L trained on Dataset 2 for 150 epochs, showing high precision and recall in test scenarios.
Comparative Findings: YOLOv8 models demonstrated superior performance in real-world video tests, detecting more brittle stars accurately.

Ocean Exploration and Research

Model Comparison: E-JUST

Goal: Develop a robust object detection system using YOLO (You Only Look Once) models to identify brittle stars in underwater images, aiding in marine biodiversity research.
Approach: Utilized YOLOv8 and YOLOv10 models to assess the best balance of speed and accuracy.
Two dataset versions:

- **Version 1**: Single class (Brittle Star) for focused detection.
- **Version 2**: Two classes (Brittle Star, Not Brittle Star) to improve model's ability to distinguish brittle stars from other organisms.

Architecture	Precision	Recall	F1-Score
YOLOv8n - 200 epochs	0.57	0.62	0.60
YOLOv8s - 100 epochs	0.59	0.63	0.61
YOLOv8s - 100 epochs - rotate ±15°	0.57354	0.62	0.59
YOLOv8m - 100 epochs	0.59	0.64	0.61
YOLOv8x - 100 epochs	0.57	0.63	0.62
YOLOv10L - 100 epochs	0.57	0.65	0.61
YOLOv10B - 100 epochs	0.60	0.62	0.61

Table 3: Comparison of the models trained on Dataset 1

Architecture	Precision	Recall	F1-Score
YOLOv8s - 100 epochs - scale 0.3	0.65	0.62	0.64
YOLOv8s - 100 epochs - scale 0.8	0.62	0.64	0.63
YOLOv8m - 100 epochs	0.67	0.67	0.67
YOLOv8m - 150 epochs	0.74	0.64	0.68
YOLOv8L - 100 epochs	0.67	0.65	0.66
YOLOv8L - 150 epochs	0.69	0.65	0.67
YOLOv8L - 200 epochs	0.73	0.67	0.70

Table 4: Comparison of the models trained on Dataset 2

Ocean Exploration and Research

Composite Model Approach: Jesuit High School

Objective: Develop a system for detecting and tracking brittle stars in underwater video to assist marine biology research.

System Components:**YOLOv6** for real-time detection, optimized for speed and smaller targets.

Faster R-CNN for high-precision detection, useful in complex underwater scenes.

OpenCV CSRT Tracker for continuous tracking without re-running detections each frame.

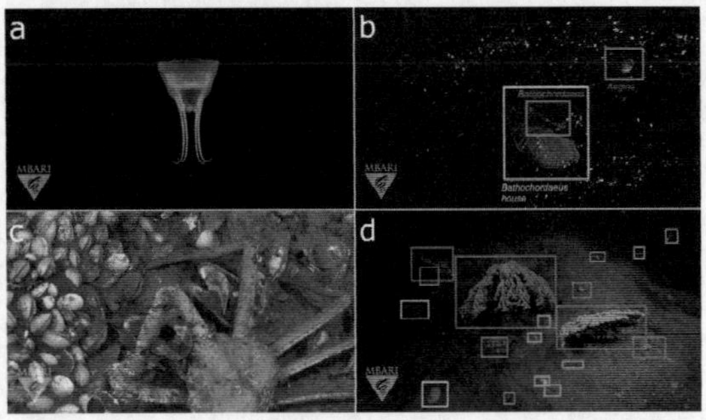

Ocean Exploration and Research

Composite Model Approach: Jesuit High School

YOLOv6:Trained on underwater data from FathomNet, optimized for small, 640x640 image regions.

- Fast processing suited for initial, coarse detections.

Faster R-CNN: Provides precise detection, trained on FathomNet specifically for brittle stars.

- Complements YOLOv6 for complex scenes by identifying details YOLOv6 might miss.

Fusion of Outputs: Detections from both models are combined based on Intersection over Union (IoU) threshold (0.45), reducing redundancy by merging overlapping detections into single boxes.

Ocean Exploration and Research

Composite Model Approach: Jesuit High School

Tracking: OpenCV's CSRT tracker updates the positions of detected brittle stars frame-by-frame, reducing computation.
Every 30 frames, YOLOv6 and Faster R-CNN run again to refresh tracking.

System Workflow:
Step 1: Detect brittle stars on the initial frame with both YOLOv6 and Faster R-CNN, merge boxes using IoU.
Step 2: Track across frames using CSRT.
Step 3: Repeat detection every 30 frames to capture new objects or scene changes.

Outputs: Annotated video and XLS file with bounding box data for further analysis.

Ocean Exploration and Research

Pre-Training Approaches: NTU Mecatron

Pretraining Phase

- **Objective**: Adapt the YOLOv10m model for underwater brittle star detection in low-visibility environments.
- **Strategy**: Leverage extensive pretraining on a large, diverse underwater dataset (~25,000 images across 8 marine classes) to familiarize the model with underwater conditions.

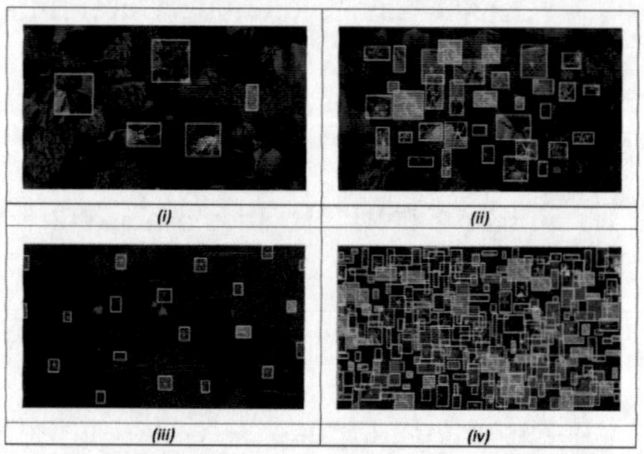

Figure 1: Four samples of random synthetic images with auto-generated bounding boxes only for brittle stars, generated by Unity Simulator.

Ocean Exploration and Research

Pre-Training Approaches: NTU Mecatron

Fine-Tuning Phase

- **Dataset**: Specialized fine-tuning dataset (~5,000 images) focused exclusively on brittle stars, combining real (425) and synthetic (2,100) images for increased accuracy.
- **Process**: Initial training on the large dataset followed by targeted fine-tuning, optimizing the model's detection accuracy for the brittle stars in murky water.

Model	Params (M)	FLOPs (G)	$mAP^{val}_{50\text{-}95}$	Latency (ms)	Latency-forward (ms)
YOLOv6-3.0-N	4.7	11.4	37.0	2.69	**1.76**
Gold-YOLO-N	5.6	12.1	**39.6**	2.92	1.82
YOLOv8-N	3.2	8.7	37.3	6.16	1.77
YOLOv10-N	**2.3**	**6.7**	39.5	**1.84**	1.79
YOLOv6-3.0-S	18.5	45.3	44.3	3.42	2.35
Gold-YOLO-S	21.5	46.0	45.4	3.82	2.73
YOLOv8-S	11.2	28.6	44.9	7.07	**2.33**
YOLOv10-S	**7.2**	**21.6**	**46.8**	**2.49**	2.39
RT-DETR-R18	20.0	60.0	46.5	**4.58**	**4.49**
YOLOv6-3.0-M	34.9	85.8	49.1	5.63	4.56
Gold-YOLO-M	41.3	87.5	49.8	6.38	5.45
YOLOv8-M	25.9	78.9	50.6	9.50	5.09
YOLOv10-M	**15.4**	**59.1**	**51.3**	4.74	4.63

Figure 2: Comparison of different generations and sizes of YOLO models. (Source: Ultralytics)

Ocean Exploration and Research

3. Winning Designs

3. Medford High School

Model Choice: Selected YOLO for its speed and adaptability, allowing accurate detection with minimal data requirements.
Training: Trained on only 3 annotated images, achieving robust identification of approximately 200 brittle stars per frame in video analysis.
Deployment: Python-based script draws bounding boxes and saves detection data in CSV format, ensuring compatibility across platforms.
Interface: Simple command-line interface (CLI) enables easy deployment, with support for custom paths and adjustable parameters for input video, model, and output settings.

Ocean Exploration and Research

2. M.I.A Robotics

Iterative Agile Process: Enhanced adaptability and continuous improvement via feedback loops, rapid prototyping, and team collaboration.
Dataset: Collected diverse marine life images, including augmentation (e.g., blurring, flipping) to improve model robustness. Split data into 85% training, 10% validation, and 5% testing.

Ocean Exploration and Research

2. M.I.A Robotics

Model Choice: Selected RE DETR for efficient multi-scale detection, IoU-aware query selection, and adjustable inference speed, optimizing for real-time accuracy and performance.

Training: 100 epochs on a T4 GPU, achieving high mean Average Precision (mAP).

Deployment: Supports varied inputs (live video, stored media), with an easy-to-use CLI and optional Web Interface.

Ocean Exploration and Research

2. M.I.A Robotics

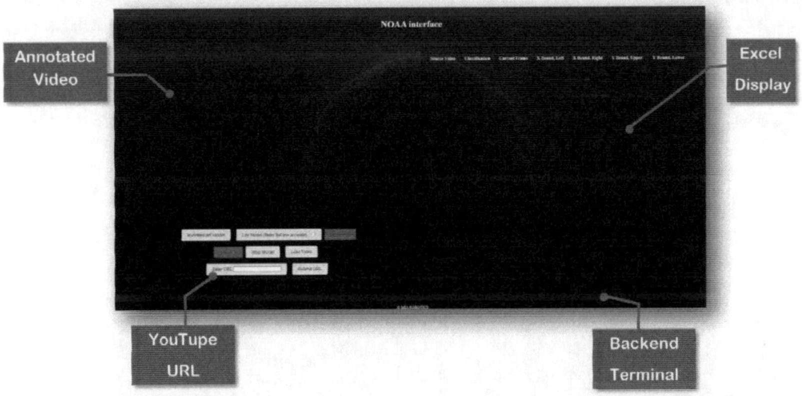

Ocean Exploration and Research

2. M.I.A Robotics

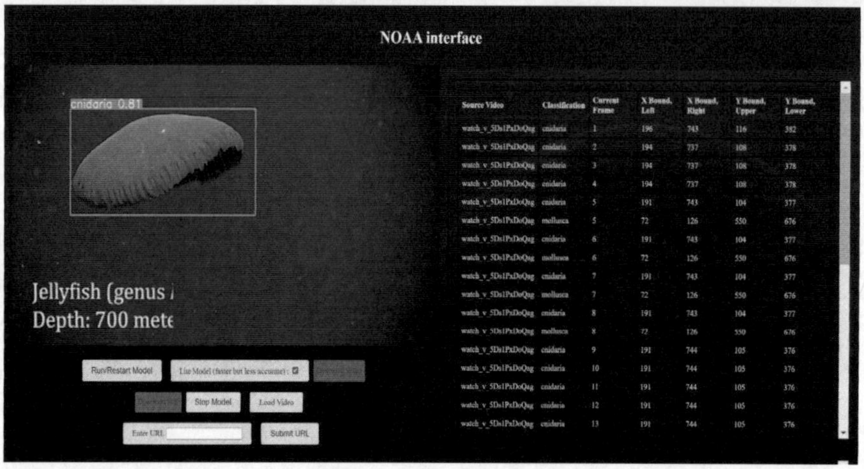

Ocean Exploration and Research

The Winning Design: Introduction

- I am the CTO of the Night Owls MATE ROV Team
- Our team is from FAU High in Boca Raton
- This is our fourth year competing in the MATE ROV competition, but our first year attempting Ocean Exploration Video Challenge.

Ocean Exploration and Research

The Problem

Too many objects to label manually

- Roughly 200 instances per frame across 1800 frames!

A large fraction of the objects are small and distant from the camera

Limited training data

- Adjacent frames are very similar

Cropped example of model output

Ocean Exploration and Research

Our Approach

- Decided to leverage the yolov8 family of pretrained models with fine tuning on a mostly synthetic data set
- Our data set was created primarily by using Segment Anything 2 to track manual a small number of manual annotations through the training video, followed by manual data cleanup.
- Using this training data recipe, we trained a series of models with a variety of hyperparameters

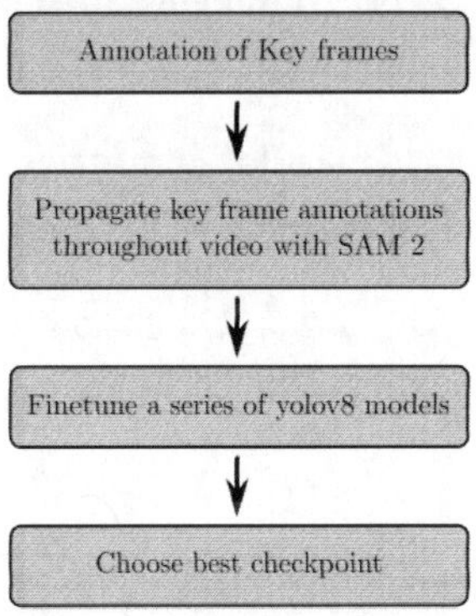

Ocean Exploration and Research

Hyperparameter Optimization

Training resolution trade off

- Affects the model's ability to recognize distant objects

Model size trade off

- Larger models improve accuracy at the cost of over fitting

Training with "dropout"

- Reduces overfitting, but increases training duration

Ocean Exploration and Research

Choosing the Best Model

After running a series of experiments with different hyper parameters, we wanted a quantitative way to compare model checkpoints

We calculated metrics for temporal coherence, and box loss with respect to a small dataset (n = 4) of manually annotated frames

The model checkpoints with the best metrics were then subjectively compared by looking at differences in annotation placement

Ocean Exploration and Research

3. Takeaways

Key Takeaways

- Simulated Imagery and Pretraining are Promising
- Conventional CNN-based Architectures are still Excellent Choices in the Era of Vision-based Transformers
- Significant results can be achieved with very few annotated images
- Complexity is not equivalent to Capability
- While YOLO was effective, there were very few overlapping organisms. YOLO's IOU-based elimination may not always work.

Ocean Exploration and Research

PROJECT ROVe

Empowering Educators, Creating Leaders

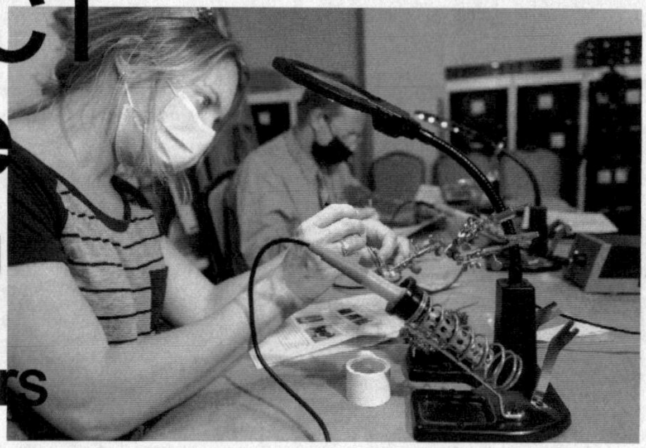

Andrea Schmuttermair, *ROV Education Coordinator*
National Marine Sanctuary Foundation
NOAA Teacher at Sea Alumni Association (TASAA)

Andrea Schmuttermair

ROV Education Coordinator, NOAA Teacher at Sea Alum (2012 & 2015

Exploring GEOMAR Oceanographic Research Centre and their ROVs in Kiel, Germany

First year building ROVs at Colorado STEM Academy

TAS 2012, Gulf of Mexico, Groundfish Survey

TAS 2015, Gulf of Alaska (Kodiak), Walleye Pollock Survey

ROVs in the Classroom
From Colorado to Germany

Students test their ROVs at a local dive shop as part of a unit learning how people explore the ocean.

Students visit GEOMAR Helmholtz Centre for Ocean Research in Kiel, Germany to test their ROVs and learn about the variety of marine technology being used today.

PILOTING PROJECT

Creating engaging professional development courses for educators

Project ROVe gives educators the opportunity to:

- ENGAGE with each other and experts in the field
- EDUCATE themselves in an interesting and timely topic
- Confidently develop curriculum for their students to EXPLORE the environment through marine technology

Why ROVs?

Remotely Operated Vehicles (ROVs) are ideal tools for engaging students in real-world problem solving.

Designing and building ROVs requires students to experiment, innovate, collaborate, think critically, and reflect—the skills we want to develop in our leaders of tomorrow.

When we build capacity in our teachers, they gain the skills and confidence to bring new topics of study—like ROVs—to their students, having an impact on generations to come.

"When I was first introduced to ROVs, probably five or six years ago, I thought... 'Oh, my gosh, this would just be amazing.' And then as I started to read, I just got completely overwhelmed and felt like there's no way. I can't. I don't have the electronics background, I don't have the mechanical background, I'm not an engineer. I genuinely felt it was just too much, way too much. And so when this opportunity came around, the chance to do ROVs as a cohort to work through this together, and to be led, I thought, 'Okay, here's my chance, I'm gonna try this.'"

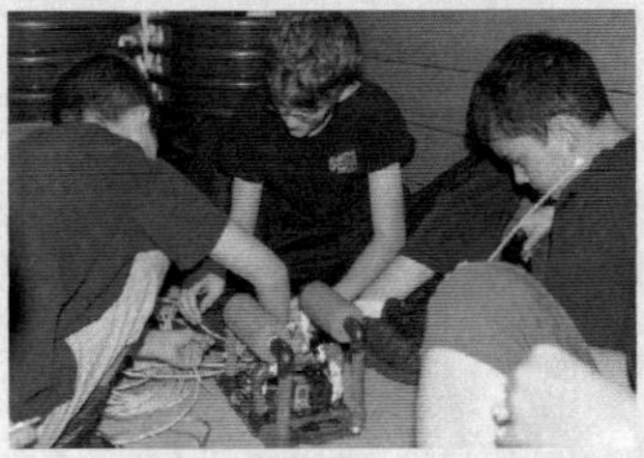

PROJECT ROV Courses

Course Syllabus

Introduction to ROVs

- Online course
- Explores broad ROV-related topics
- Activities, virtual meetings, discussions, assignments

Course Syllabus

Design and Build

- In-person workshop
- Build a MATE Pufferfish ROV
- Create a classroom implementation plan
- * Introduction to ROVs is a pre-requisite for this course.

Marine Technology Courses

Introduction to Sensors with Micr

- Geared towards teachers in any subject area who want to learn how to use sensors with their students to collect and analyze data

Course Syllabus

MultiSensors and Data Loggers

Public Sensors

- Geared towards teachers who want to build a multi-sensor to measure temperature, light and distance
- Taught in conjunction with Dr. Sasha Seroy and Dr. Robert Levine from the University of Washington School of Oceanography

Course Syllabus

Marine Technology Courses

Oceanography for Everyone: Op

- This course will take educators through the build, calibration, and testing of a low-cost CTD. Educators will bring home an OpenCTD kit to build with their students and design projects using the CTD to collect data in their local area.

Coming Soon

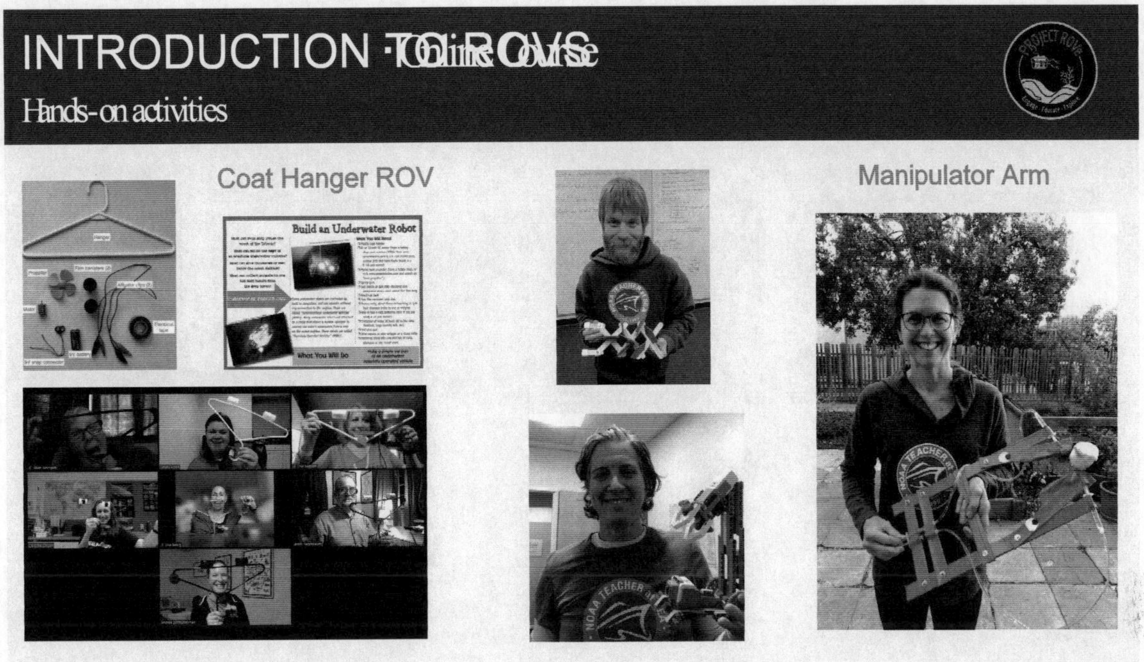
INTRODUCTION TO ROVS
Online Course
Hands-on activities
Coat Hanger ROV
Build an Underwater Robot
What You Will Do
Manipulator Arm

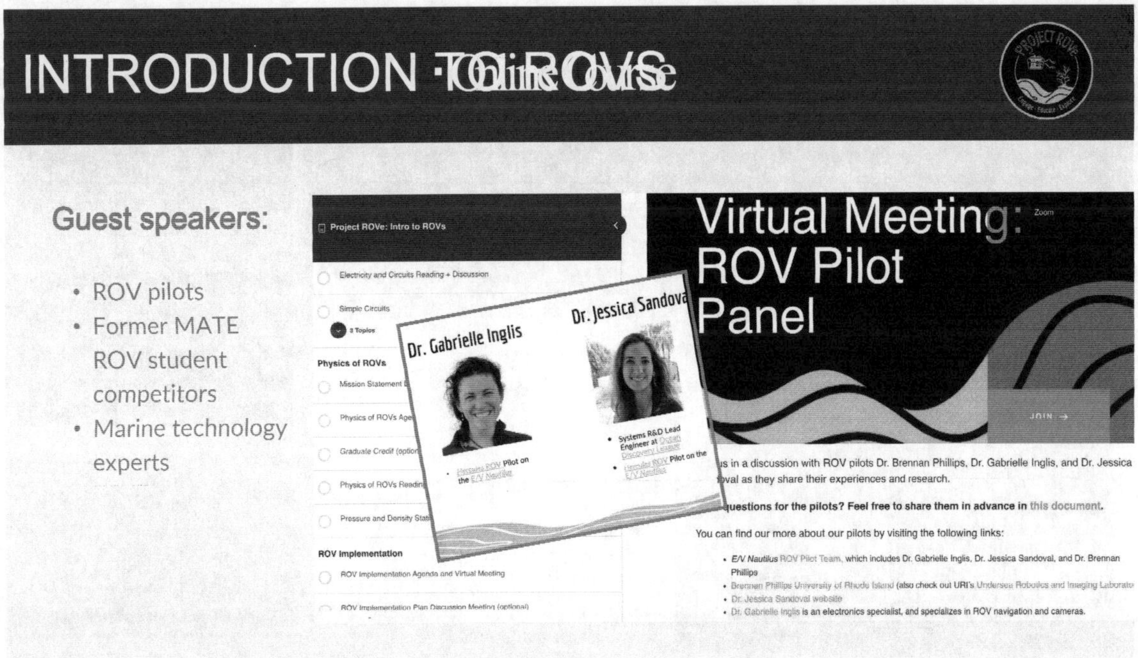
INTRODUCTION TO ROVS
Online Course
Guest speakers:
• ROV pilots
• Former MATE ROV student competitors
• Marine technology experts
Project ROVe: Intro to ROVs
Electricity and Circuits Reading + Discussion
Simple Circuits
3 Topics
Physics of ROVs
Mission Statement
Physics of ROVs
Graduate Credit
Physics of ROVs Reading
Pressure and Density
ROV Implementation
ROV Implementation Agenda and Virtual Meeting
Dr. Gabrielle Inglis
Dr. Jessica Sandoval
Virtual Meeting: ROV Pilot Panel
Zoom
JOIN →
You can find our more about our pilots by visiting the following links:

Design and Build ROV Worksh

- Monitor National Marine Sanctuary in Newport News, VA (2022)
- Gray's Reef National Marine Sanctuary in Savannah, GA (2023)
- Olympic Coast National Marine Sanctuary in Forks, WA (2024)

Place-based exploration for educators leads to local stewardship and student projects

PROJECT ROVe
OLYMPIC COAST
National Marine Sanctuary

Project ROVe By the Numbers
2022-2024

1,300+
Students actively involved in ROV projects

50
Educators participating in Project ROVe

19 STATES
EDUCATORS FROM 19 STATES PLUS BERMUDA AND AMERICAN SAMOA

PROJECT IMPLEMENTATION
70%
Teachers implementing ROV projects by Spring 2024

PROJECTS FUNDED
7
Projects funded by the TASAA/MATE/NOAA OE

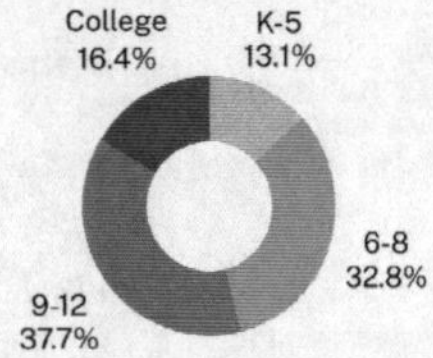

GRADE LEVEL
EDUCATORS REACHING STUDENTS ACROSS ALL GRADES

HEATHER WILLIAMSON
FIRST GRADE TEACHER AND EDUCATIONAL TECHNOLOGY F

28 STUDENTS — ROV Class Project (2022-23)
1st grade
Haymarket Elementary
Haymarket, VA

24 STUDENTS — ROV Club (2023-24)
4th-5th grade
Kyrene de la Mariposa Elementary
Tempe, AZ

2-year impact

2 states — AZ, VA

6 schools

250 students

125 STUDENTS — Marine Science and ROV Unit (2023-24)
Grades 6-8
Kyrene Middle School
Tempe, AZ

75 STUDENTS — Marine Science and ROV Unit (2023-24)
Grades 6-8
Altadeña Middle School
Ahwatukee, AZ

*Heather received an Ocean Odyssey Grant for her ROV work in Arizona in 2024.

When we build capacity in our educato

"Project ROVe gave me the ***confidence*** *I needed to build ROVs from start to completion with my students. I had a blast learning the different concepts in an excellent* ***community of practice*** *I needed the hands on workshop to make this happen for our Bermuda students this summer." Kaitlin N., 5-12th grade teacher*

"It was fantastic, and easily one of the most ***fun and engaging*** *PDs I have ever done." Caroline S. HS Chemistry teacher*

"The best professional development course I've ever taken, Project ROVe provides a hands-on introduction to underwater robotics that revealed a ***new world*** *for my students and I to explore." Jason M., PK-adult educator*

"I can't thank you enough for all you did to help make this happen. I really appreciate the training and support along the way. This was a wonderful project and the sight of 5th graders doing ***engineering in real time*** *will stay with me." Jacob T., 5th grade teacher*

...our students fly, too!

COLLABORATION AND SUPPORT

This map shows the locations past Project ROVe participants.

Participants are encouraged to connect with each other for support and collaboration.

Click the map to take you to the website and find out who is in each region!

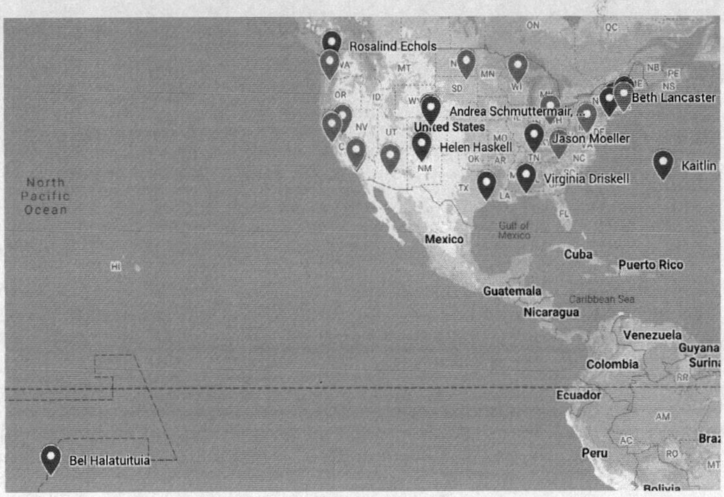

OCEANEERING®

Systems Engineering for Subsea Systems

Gary Wieboldt, ESEP
Chief Systems Engineer
Oceaneering AdTech/OTECH

13 December 2024
Underwater Intervention, New Orleans, LA

Agenda

1. A bit about me
2. SE introduction
3. Tenets and the Vee
    1. Why
    2. What
    3. How
    4. How well
4. Case study: Why requirements are important
5. Summary

A bit about me…

- 42+ years Engineering experience
- INCOSE ESEP certification
- BSEE, MS Computer Science, MS Systems Engineering
- Developed subsea systems (ROV, seafloor systems)
- Developed the first successful 20KFSW ROV SM optical fiber Telemetry system (Gemini-1988)
- Involved with cradle-to-grave lifecycle, have wealth of offshore 'experiences'

"…. going offshore regularly keeps you humble…"

A bit about me 3 OCEANEERING

What is Systems Engineering?: a SE geek view

- Systems Engineering is a transdisciplinary and integrative approach to enable the successful realization, use, and retirement of engineered systems, using systems principles and concepts, and scientific, technological, and management methods.[1]

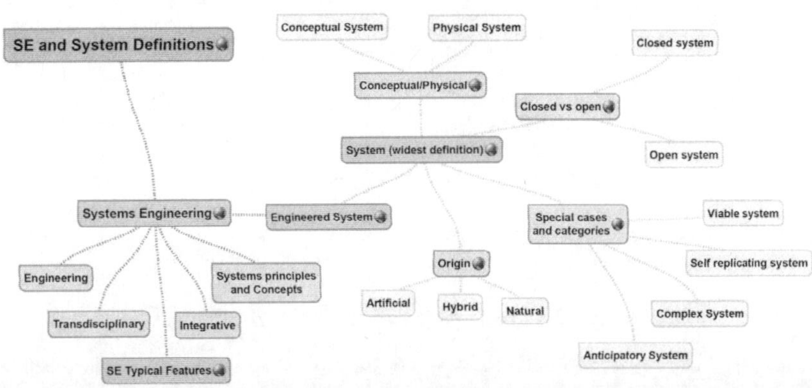

SE Introduction 4 OCEANEERING

What is SE? 1877 Marine Architects View[2]

THEORETICAL NAVAL ARCHITECTURE.

INTRODUCTION.

The work of the Naval Architect is to design a floating structure, stable, seaworthy, and strong; capable of being propelled with facility either by the wind, by steam, or some other motive power, and adapted for the particular service in which it is to be employed, whether as a fighting ship or for carrying merchandise or passengers. In every case certain qualities, viz., stability, sea-worthiness, strength, speed, stowage, and accommodation, have to be fulfilled in a varying ratio according to the purpose for which the ship is intended. In a ship of war certain other qualities have to be obtained, such as power of carrying and fighting guns, and sometimes invulnerability; these too being provided in different ratios of importance according to the duties the ship has specially to perform. It is the art of securing these qualities, and apportioning them in the several degrees necessary to secure the maximum efficiency of the ship in her peculiar vocation, that constitutes the functions of the Naval Architect, and renders his duties among the most difficult and onerous that fall into the hands of man to perform; requiring, as they do, a knowledge of some of the most complex and abstruse physical laws, and a wise discrimination in applying experimental data.

The development of the science of Naval Construction has

Varying ratio

Ratio of importance

"Varying ratio according to purpose"[3]

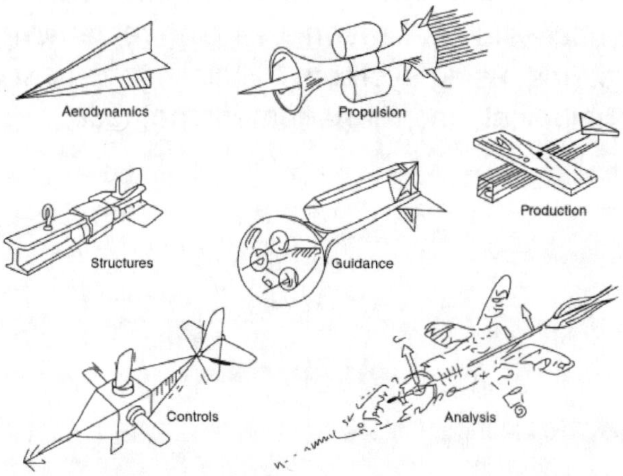

If an airplane was designed *only* by a specialty…

"Ratio of importance"

How important was MCAS implementation on the Max 8?

Successful systems require balance...

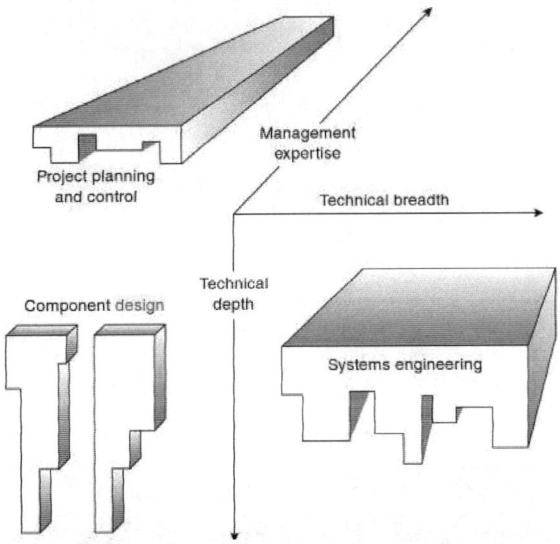

... of cost, schedule and technical performance

Because if you need to make changes late…

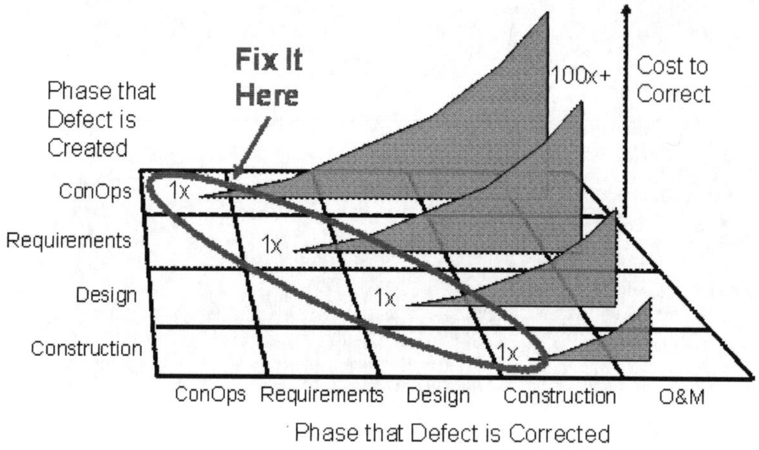

… it can cost you BIG!

Using SE

You need to follow the plan:

1. Why
2. What
3. How
4. How well

The Tenets of Systems Engineering

For the purposed of this talk: we are discussing classical 'Vee' SE and not DoDAF, MBSE, SySML etc

The *Why*:

Why does the customer need the System?

1. You enter the development from one of 2 perspectives
   1. Requirements Push:[Customer: pick up black box of downed airliner]
   2. Technology Pull: [NASA- James Webb Space Telescope]
2. Risk and technology proposition(s) varies with each

It is important to understand the **Why**. It drives the process. It drives the outcome.

The *What*

1. What does it HAVE to do? ***<u>Require</u>***ments
   1. Size, Weight and Power (SWaP)
   2. Depth(more on this later!)
   3. Functions: (Verb Noun construct-functional analysis):
      1. Process (V) potatoes (N)
      2. Shed Water
      3. Move cable
      4. Rectify power

A list of agreed to bounded parametric requirements, widely communicated is key
Even if they are at the top level

The *How*

This is the engineering detail:

Given:
(What)Develop a 12.5 VDC (+-10%) power supply delivering up to 10 Amps
(How)Use a 7812 IC driving 2ea 2N2035 pass transistors in parallel…

KNOW THE ***WHAT*** BEFORE YOU FIGURE OUT THE ***HOW***

Fundamental paradigm
Fundamental failure in many systems

Do not start at this step!...

The *How well*

Verification that you met the requirements:

IADTC: Inspection, Analysis, Demonstration, Test and Certification
(listed in approximately least costly to most costly)

Inspection:
The nondestructive examination of a product or system using one or more of the five senses (visual, auditory, olfactory, tactile, taste). It may include simple physical manipulation and measurements.

Analysis:
The verification of a product or system using models, calculations and testing equipment. Analysis allows someone to make predictive statements about the typical performance of a product or system based on the confirmed test results of a sample set or by combining the outcome of individual tests to conclude something new about the product or system. It is often used to predict the breaking point or failure of a product or system by using nondestructive tests to extrapolate the failure point.

The *How Well*

Demonstration:
Manipulation of the product or system as it is intended to be used to verify that the results are as planned or expected.

Test:
Verification of a product or system using a controlled and predefined series of inputs, data, or stimuli to ensure that the product or system will produce a very specific and predefined output as specified by the requirements.

Certification:
An independent third party to assess conformance to requirements

An example

Why is it developed?
What does it do?
How does it do it?
How well did it do it?

Click for 2 min video link: Retroencabulator

The *Why*

Here at Rockwell Automation's world headquarters, research has been proceeding to develop a line of automation products **that establishes new standards for quality, technological leadership, and operating excellence.** With customer success as our primary focus, work has been proceeding on the crudely conceived idea of an instrument that would not only provide inverse reactive current for use in unilateral phase detractors, but would also be capable of automatically synchronizing cardinal grammeters. Such an instrument comprised of Dodge gears and bearings, Reliance Electric motors, Allen-Bradley controls, and all monitored by Rockwell Software is Rockwell Automation's retro-encabulator.
Now basically the only new principle involved is that instead of power being generated by the relative motion of conductors and fluxes, it's produced by the modial interaction of magneto reluctance and capacitive duractance. The original machine had a base plate of pre-famulated amulite surmounted by a malleable logarithmic casing in such a way that the two spurving bearings ran in a direct line with the panametric fan.
The line-up consisted simply of six hydrocoptic marzul vanes so fitted to the ambaphascient lunar wain shaft that side fumbling was effectively prevented. The main winding was of the normal lotus-odeltoid type placed in panendurmic semi-bulloid slots of the stator, every seventh conductor being connected by a non-reversible tremmy pipe to the differential girdle spring on the up-end of the grammeters.
Moreover, whenever fluorescent square motion is required, it may also be employed in conjunction with the drawn reciprocation dingle arm, to reduce sinusoidal depleneration.
The retro-encabulator has now reached a high level of development, and it's being successfully used in the operation of Milford trunnions. It's available soon, wherever Rockwell Automation products are being sold.

An example 17 OCEANEERING

The *What*

Here at Rockwell Automation's world headquarters, research has been proceeding to develop a line of automation products that establishes new standards for quality, technological leadership, and operating excellence. With customer success as our primary focus, work has been proceeding on the crudely conceived idea of an instrument that would not only **provide inverse reactive current** for use in unilateral phase detractors, but would also be capable of **automatically synchronizing cardinal grammeters.** Such an instrument comprised of Dodge gears and bearings, Reliance Electric motors, Allen-Bradley controls, and all monitored by Rockwell Software is Rockwell Automation's retro-encabulator.
Now basically the only new principle involved is that instead of power being generated by the relative motion of conductors and fluxes, it's produced by the modial interaction of magneto reluctance and capacitive duractance. The original machine had a base plate of pre-famulated amulite surmounted by a malleable logarithmic casing in such a way that the two spurving bearings ran in a direct line with the panametric fan.
The line-up consisted simply of six hydrocoptic marzul vanes so fitted to the ambaphascient lunar wain shaft that side fumbling was effectively prevented. The main winding was of the normal lotus-odeltoid type placed in panendurmic semi-bulloid slots of the stator, every seventh conductor being connected by a non-reversible tremmy pipe to the differential girdle spring on the up-end of the grammeters.
Moreover, whenever fluorescent square motion is required, it may also be employed in conjunction with the drawn reciprocation dingle arm, to reduce sinusoidal depleneration.
The retro-encabulator has now reached a high level of development, and it's being successfully used in

You don't have to be a domain expert to listen to the Verb-noun of 'What'

An example 18 OCEANEERING

The *How*

Here at Rockwell Automation's world headquarters, research has been proceeding to develop a line of automation products that establishes new standards for quality, technological leadership, and operating excellence. With customer success as our primary focus, work has been proceeding on the crudely conceived idea of an instrument that would not only provide inverse reactive current for use in unilateral phase detractors, but would also be capable of automatically synchronizing cardinal grammeters. Such an instrument comprised of Dodge gears and bearings, Reliance Electric motors, Allen-Bradley controls, and all monitored by Rockwell Software is Rockwell Automation's retro-encabulator.

Now basically the only new principle involved is that instead of power being generated by the relative motion of conductors and fluxes, it's produced by the modial interaction of magneto reluctance and capacitive duractance. The original machine had a base plate of pre-famulated amulite surmounted by a malleable logarithmic casing in such a way that the two spurving bearings ran in a direct line with the panametric fan.

The line-up consisted simply of six hydrocoptic marzul vanes so fitted to the ambaphascient lunar wain shaft that side fumbling was effectively prevented. The main winding was of the normal lotus-odeltoid type placed in panendurmic semi-bulloid slots of the stator, every seventh conductor being connected by a non-reversible tremmy pipe to the differential girdle spring on the up-end of the grammeters.

Moreover, whenever fluorescent square motion is required, it may also be employed in conjunction with the drawn reciprocation dingle arm, to reduce sinusoidal depleneration.

The retro-encabulator has now reached a high level of development, and it's being successfully used in the operation of Milford trunnions. It's available soon, wherever Rockwell Automation products are being sold.

An example 19 OCEANEERING®

The *How well*

Here at Rockwell Automation's world headquarters, research has been proceeding to develop a line of automation products that establishes new standards for quality, technological leadership, and operating excellence. With customer success as our primary focus, work has been proceeding on the crudely conceived idea of an instrument that would not only provide inverse reactive current for use in unilateral phase detractors, but would also be capable of automatically synchronizing cardinal grammeters. Such an instrument comprised of Dodge gears and bearings, Reliance Electric motors, Allen-Bradley controls, and all monitored by Rockwell Software is Rockwell Automation's retro-encabulator.

Now basically the only new principle involved is that instead of power being generated by the relative motion of conductors and fluxes, it's produced by the modial interaction of magneto reluctance and capacitive duractance. The original machine had a base plate of pre-famulated amulite surmounted by a malleable logarithmic casing in such a way that the two spurving bearings ran in a direct line with the panametric fan.

The line-up consisted simply of six hydrocoptic marzul vanes so fitted to the ambaphascient lunar wain shaft that side fumbling was effectively prevented. The main winding was of the normal lotus-odeltoid type placed in panendurmic semi-bulloid slots of the stator, every seventh conductor being connected by a non-reversible tremmy pipe to the differential girdle spring on the up-end of the grammeters.

Moreover, whenever fluorescent square motion is required, it may also be employed in conjunction with the drawn reciprocation dingle arm, to reduce sinusoidal depleneration.

The retro-encabulator has now reached a high level of development, and **it's being successfully used in the operation of Milford trunnions.** It's available soon, wherever Rockwell Automation products are being sold.

An example 20 OCEANEERING®

Vee for victory

Why

What

How

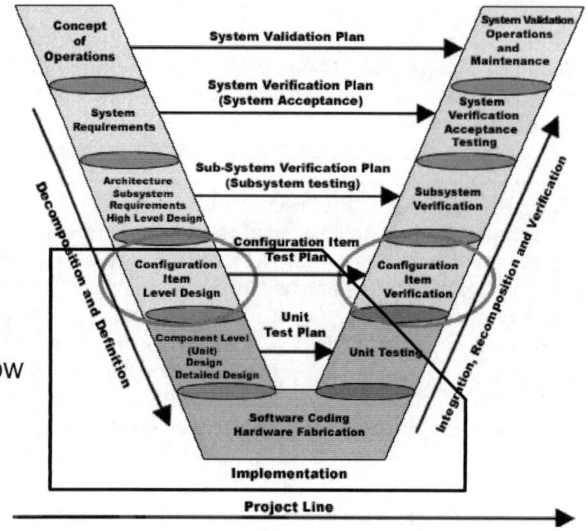

How well

Vee for victory

The Systems Engineering Vee diagram

Understand the customers mission

Develop and architect

Understand the requirements

Understand the functions

Understand the subsystems

Let the fun begin: Design!

Let the reality begin: Build

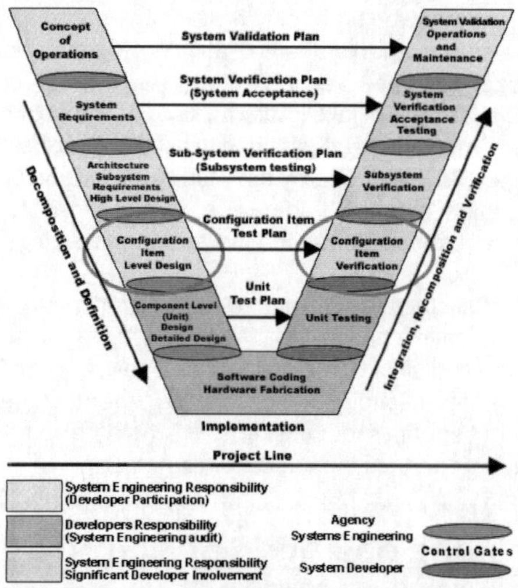

Life Cycle support

IOC

Start fielding the system

Verify (IADTC)

Verify (IADTC)

Verify (IADTC)

Let the Frustration begin: Integrate

Case study: Why requirements (What) are important

One little number can cause a BIG pain in the @$$

Case study: Why requirements (What) are important

Why
Subsea excavation too costly

This is functional analysis

What
Customer wants to test a way to excavate (V) soil (N) that is more efficient and precise

How
Use venturi system
[+other technologies]

How Well
Time on station
Excavation metrics (depth, removal rate etc)

Case study: Why requirements (What) are important

What

Customer wants to test a way to excavate (V) soil (N) that is more efficient and precise

Lots of requirements, but in particular…

"R: System Depth of up to 300MSW"

Requirements should always be S.M.A.R.T
S: Specific
M: Measurable
A: Attainable
R: Realistic
T: Testable

Case study: Why requirements are important

OCEANEERING

Case study: Why requirements (What) are important

Prime contractor generated requirements and responsible for communicating

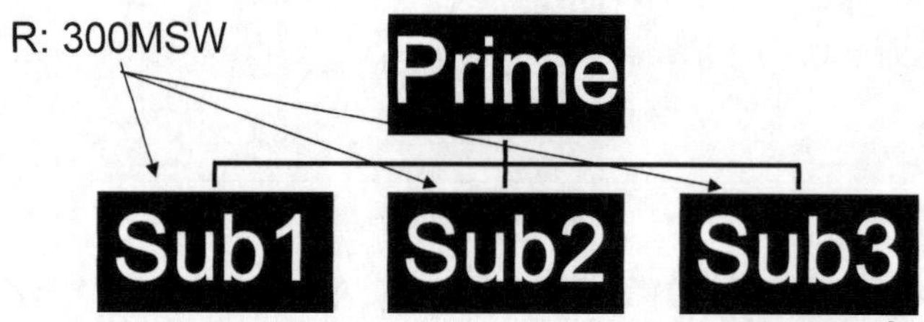

OCEANEERING

Task:
Build a fancy sensor

Task: find a patch in GOM that is:
- <=300MSW
- Has proper soil type

Case study: Why requirements are important

OCEANEERING

Case study: Why requirements (What) are important

What really happened…

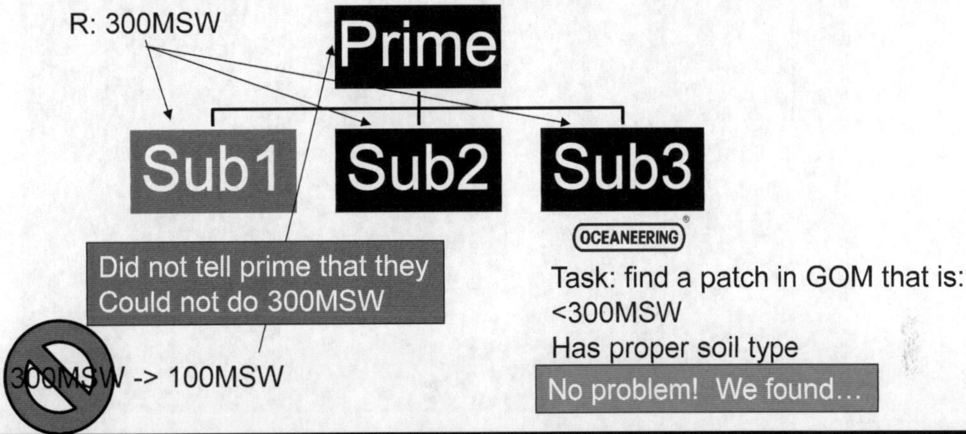

Case study: Why requirements are important

Case study: Why requirements (What) are important

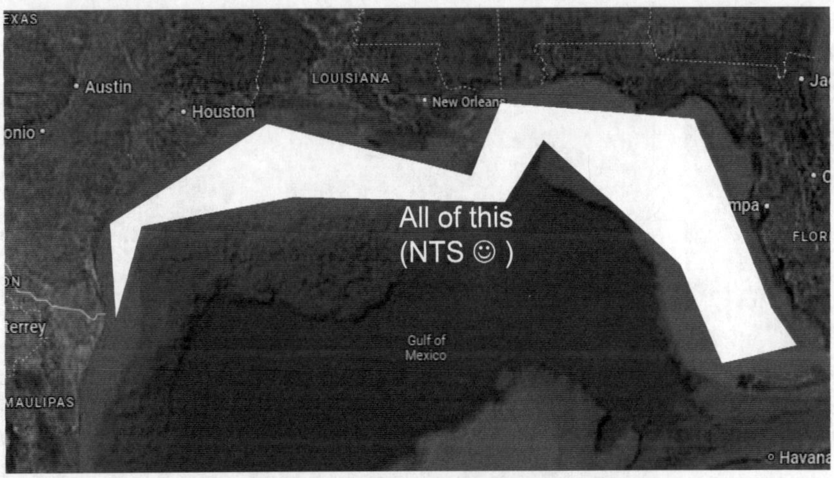

Not exact, but we had a wide area to chose from that met the requirements

Case study: Why requirements are important

Case study: Why requirements (What) are important

Late in the program we(and prime) find that the sensor was ***built*** to 100MSW! Just 2 weeks before mobilization!

So now the area that met system requirements was:

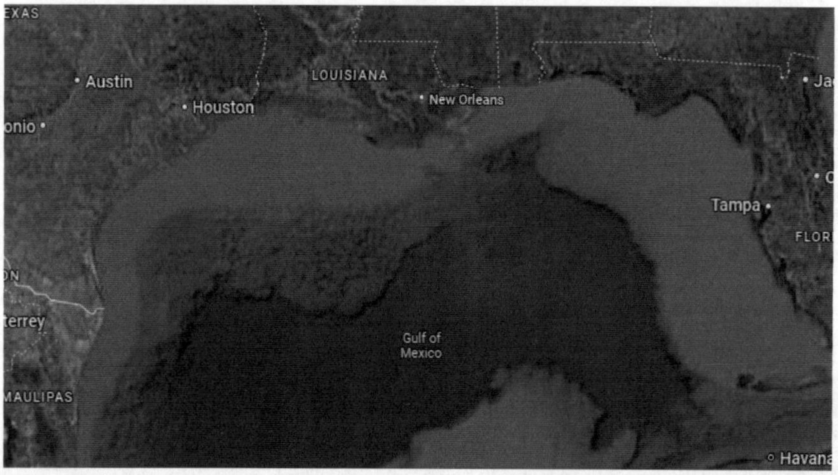

Case study: Why requirements are important

Case study: Why requirements (What) are important

What do you do? Change the earth…

Step 1: dig a 5000 ft^3 hole

Step 2: Fill with 250+ Tons of sand

Then operate the machine to…

Step 3: suck the sand out

Case study: Why requirements are important

Case study: Why requirements (What) are important

What went wrong?

- Requirements not tracked well
- Requirements not validated during design stages
- Requirements not verified during 'sell-off'

Impacts?

- Cost (++)
- Schedule (+)

Case study: Why requirements (What) are important

Before you start any project:

1. Why
2. What
3. How
4. How well

Don't start with #3!

Before you design

1. Write down the requirements
2. Freely share
3. If you cannot meet them: **communicate**- others may be involved negatively

Systems Engineering is an important part of any subsea system

Summary

- The mantra:
  - Why
  - What
  - How
  - How well
- Don't start at how
- Listen carefully, communicate fully

Thank You!

Endnotes

[1]: https://www.incose.org/about-systems-engineering/what-is-systems-engineering

[2]: Thearle, Samuel, "Theoretical Naval Architecture: a Treatise" G. P. Putnams's and Sons, New York,1877.

[3]: JHU APL MS-SE functional analysis booklet ca 1990

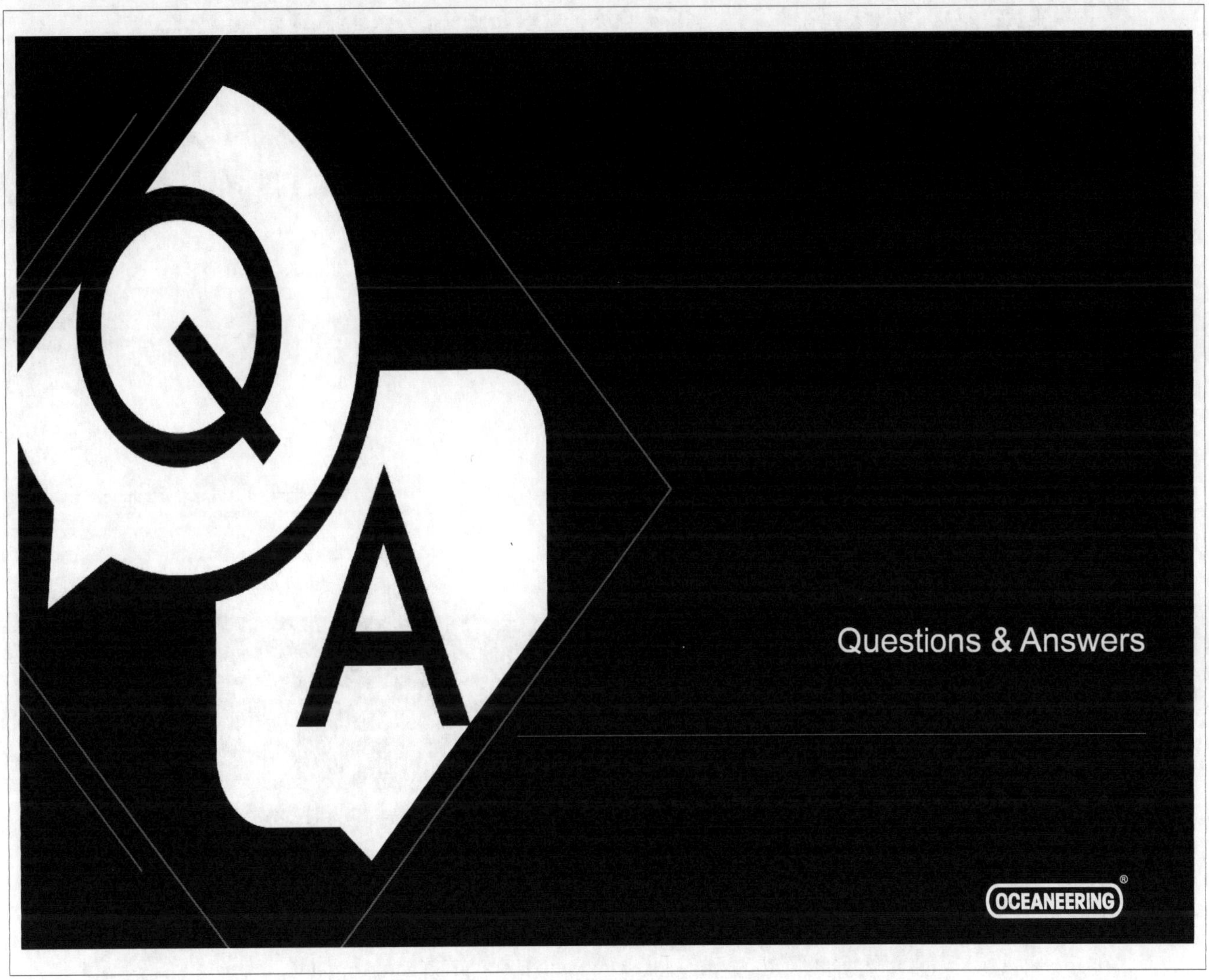
Q
A
Questions & Answers
OCEANEERING

Establishing a Community of Practice for Uncrewed Maritime Systems Data across Government, Academia, and Industry

National Centers for Environmental Information (NCEI)

Dr. Andrew Evans
Katharine Weathers

The UxS Big Data Tsunami

- Platforms and Sensors continue to improve capturing higher fidelity data
- Data Volumes have increased exponentially
- Current Data Practices present logistical challenges

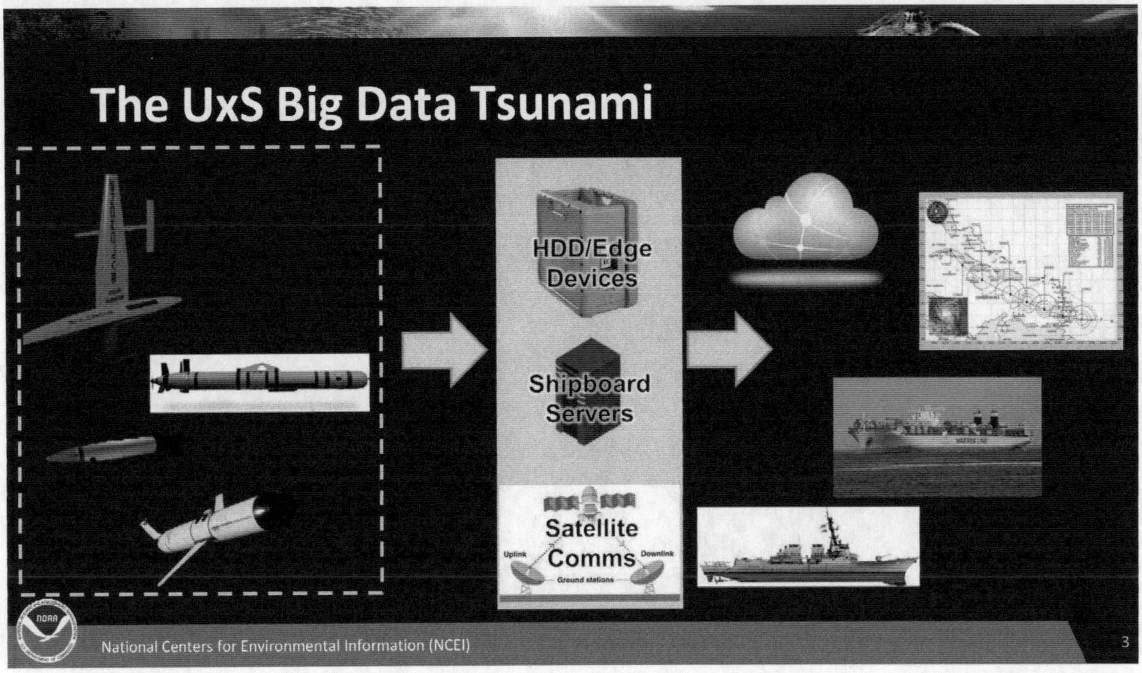
The UxS Big Data Tsunami
HDD/Edge Devices
Shipboard Servers
Satellite Comms
Uplink
Downlink
Ground stations
National Centers for Environmental Information (NCEI)
3

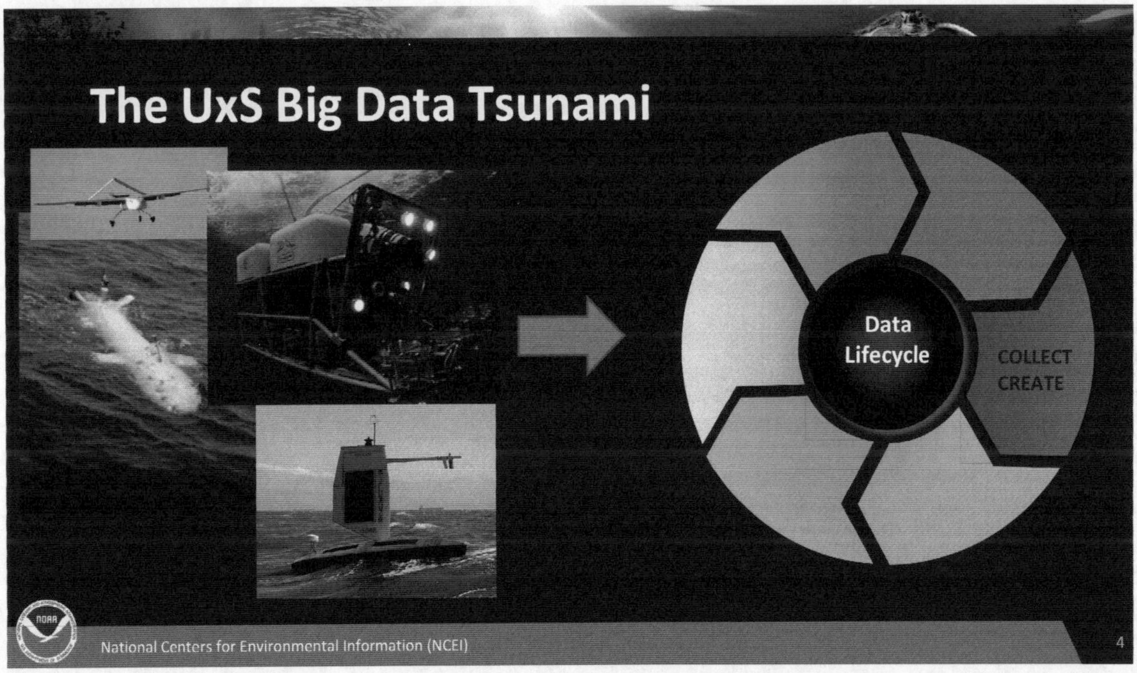
The UxS Big Data Tsunami
Data Lifecycle
COLLECT CREATE
National Centers for Environmental Information (NCEI)
4

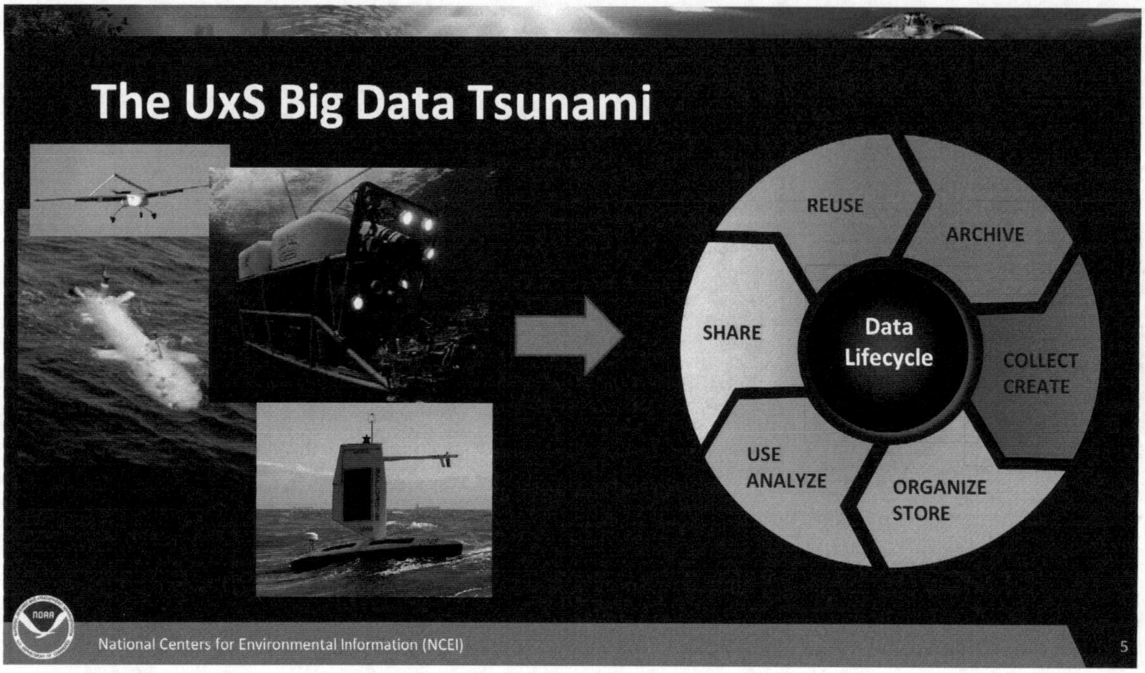
The UxS Big Data Tsunami
REUSE
ARCHIVE
SHARE
Data Lifecycle
COLLECT CREATE
USE ANALYZE
ORGANIZE STORE
National Centers for Environmental Information (NCEI)
5

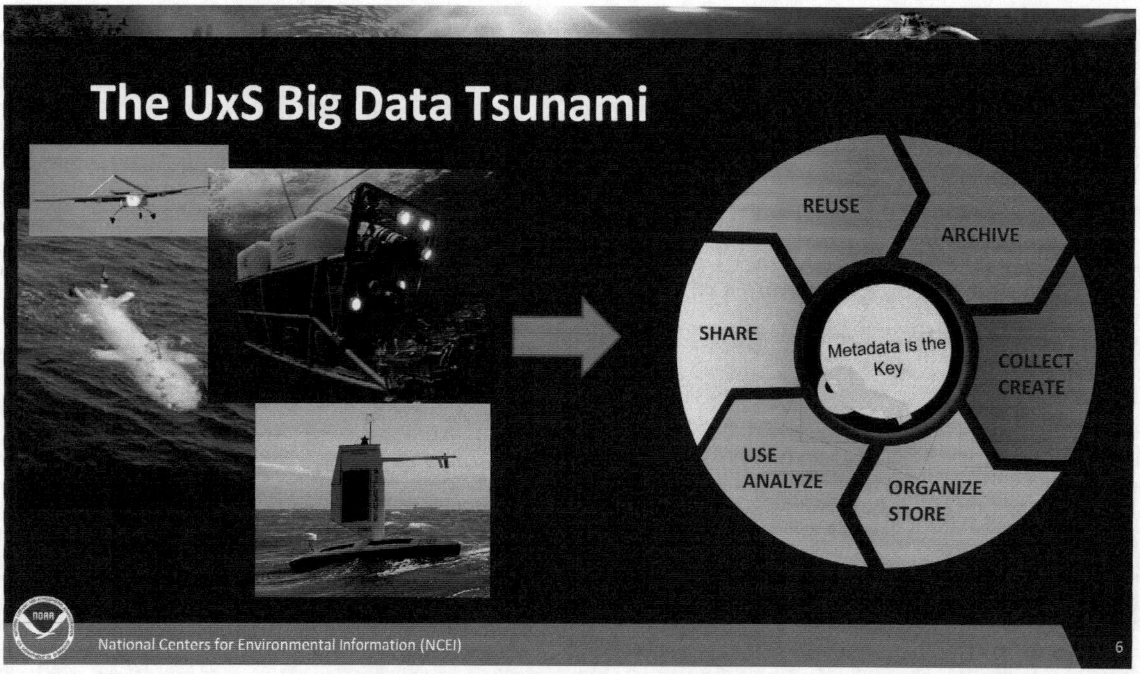
The UxS Big Data Tsunami
REUSE
ARCHIVE
SHARE
Metadata is the Key
COLLECT CREATE
USE ANALYZE
ORGANIZE STORE
National Centers for Environmental Information (NCEI)
6

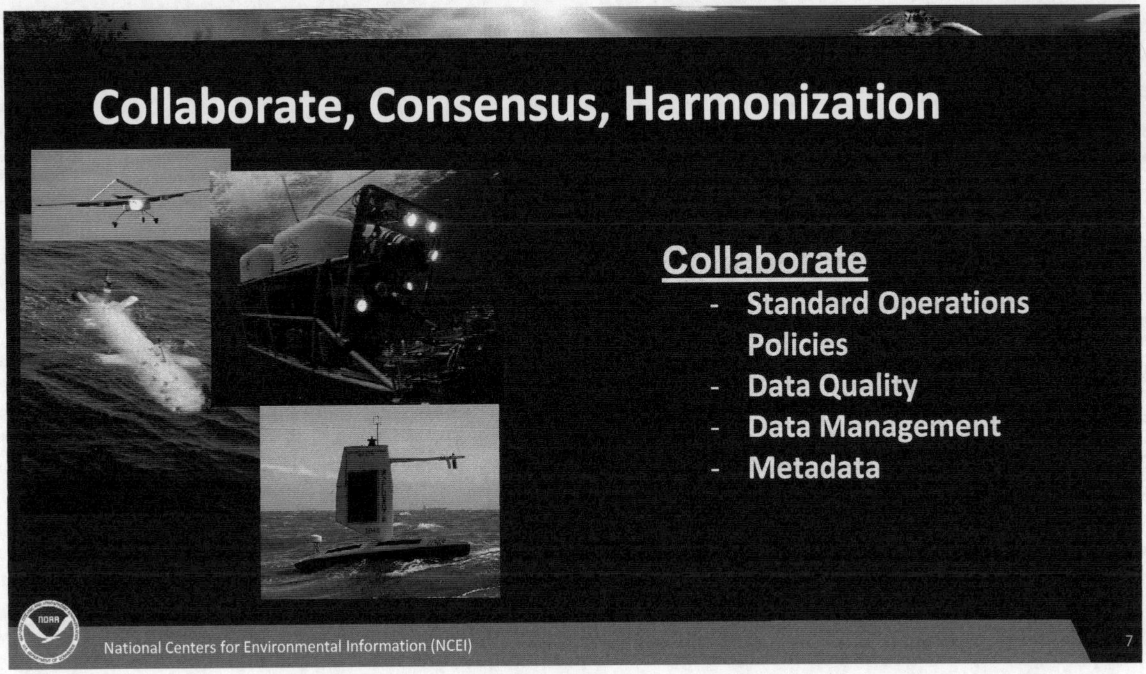
Collaborate, Consensus, Harmonization
Collaborate
- Standard Operations Policies
- Data Quality
- Data Management
- Metadata
National Centers for Environmental Information (NCEI)
7

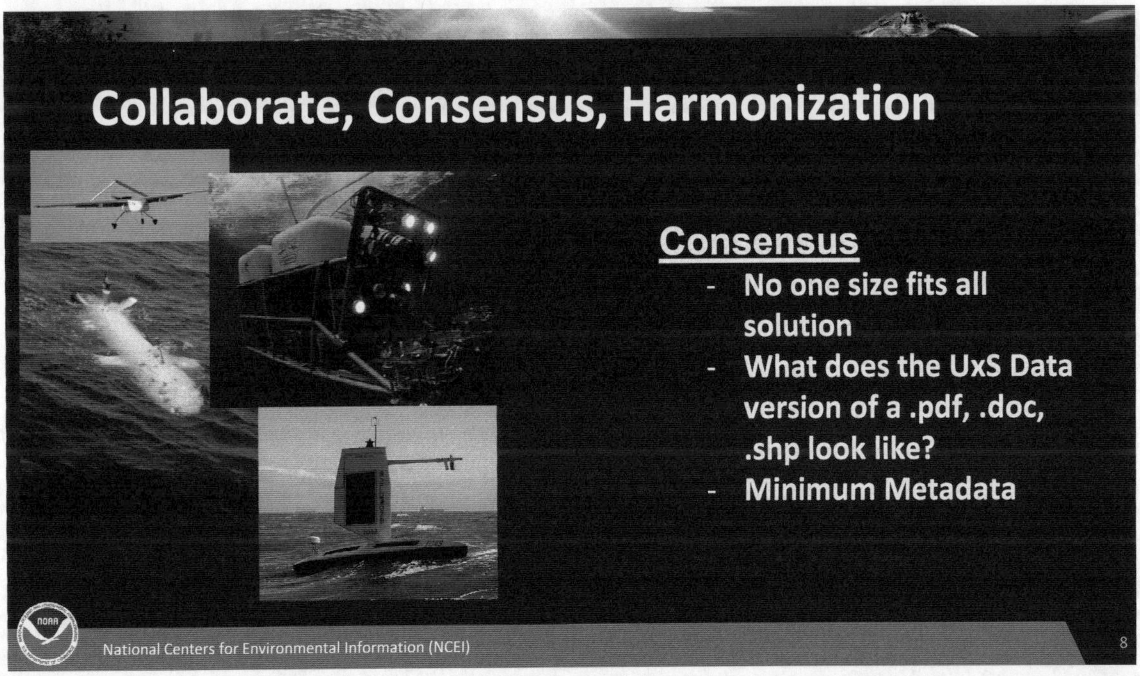
Collaborate, Consensus, Harmonization
Consensus
- No one size fits all solution
- What does the UxS Data version of a .pdf, .doc, .shp look like?
- Minimum Metadata
National Centers for Environmental Information (NCEI)
8

Collaborate, Consensus, Harmonization

Harmonization

- Keep industry in the loop
- Leverage existing technologies
- Utilize organizations such as ISO, IHO, COLREGs, MTS, AUVSI

A Starting Point

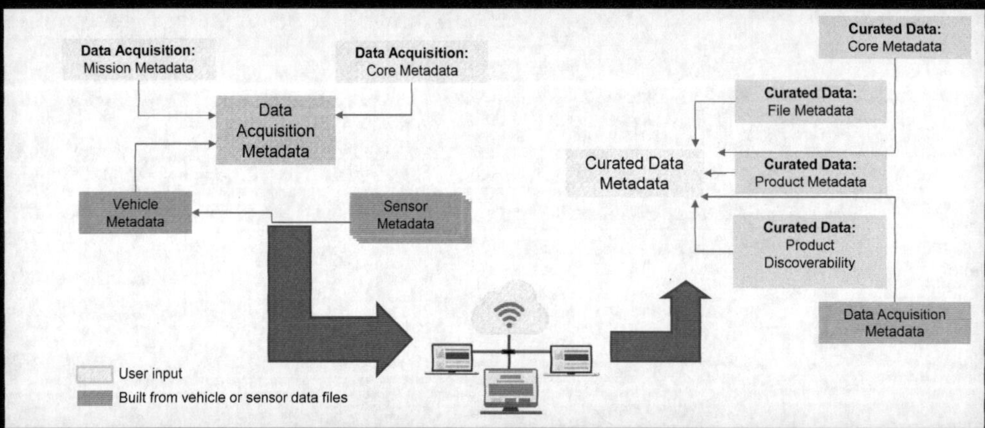

A Starting Point

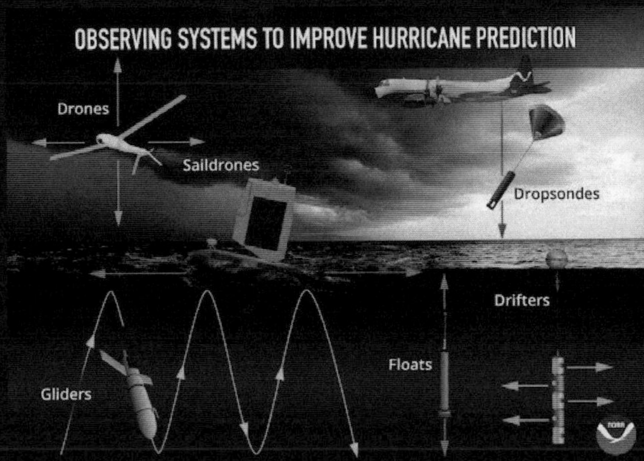

A Starting Point

- In process of building out Git Repository to share and store documentation
  - NOAA / Navy UxS XML profile
  - UAS Survey specifications from other groups both internal and external to NOAA
- Establishing Benchmarks for bathymetric surveying
- Working with ISO, DoD, and other international stakeholders
- Figuring out what gets left on the cutting room floor

Interested? Fill out our contact form to be added to our mailing list

Questions?

National Centers for Environmental Information (NCEI) 13

BSS
Baker Subsea Solutions
Redefining
"Inspection"
ROV's
Mike MacMillan
Underwater Intervention - 2024

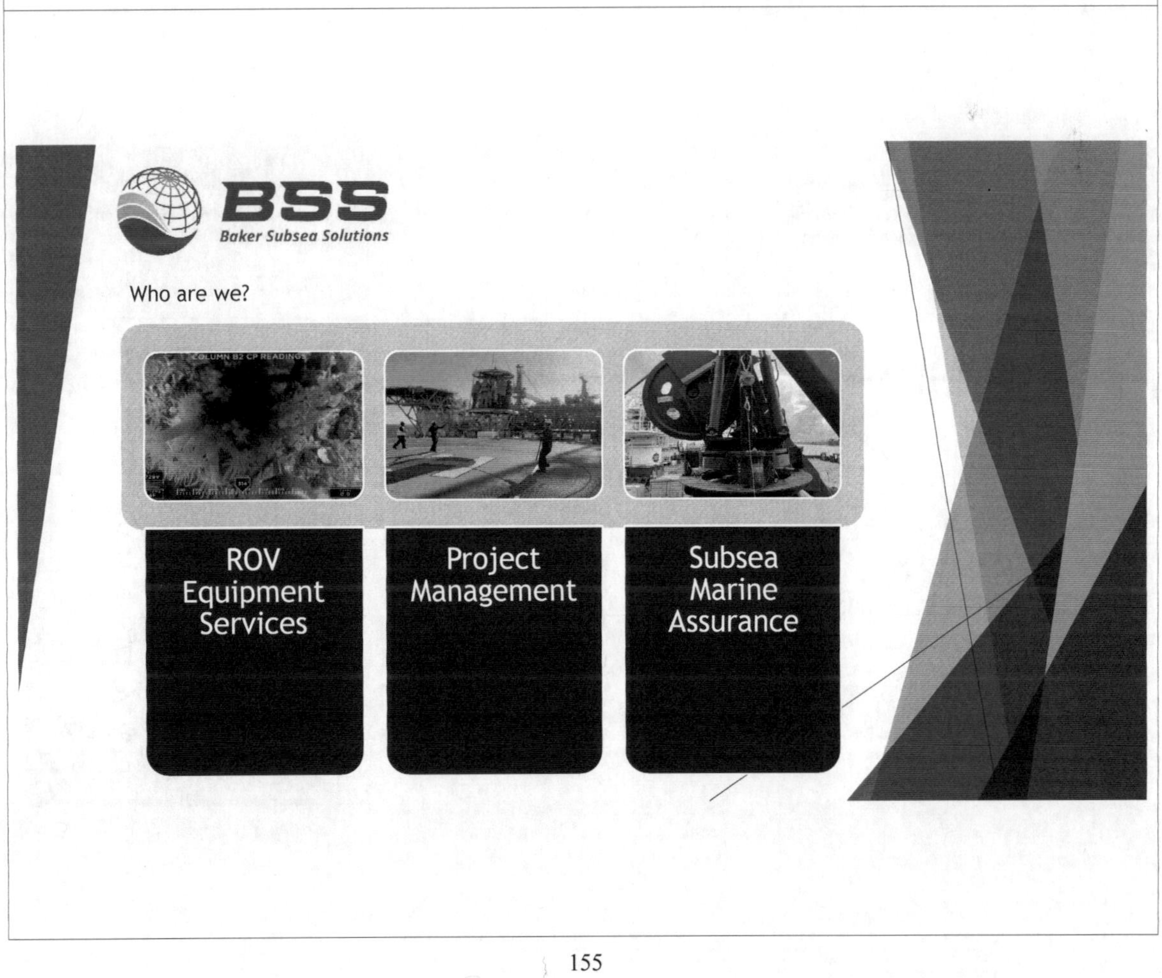
BSS
Baker Subsea Solutions
Who are we?
ROV Equipment Services
Project Management
Subsea Marine Assurance

BSS

Baker Subsea Solutions

2024 Summary

BSS

Baker Subsea Solutions

Partnerships

Greensea IQ

- Fuel Savings
- Carbon Reduction
- Maintenance Efficiency

BSS
Baker Subsea Solutions

Partnerships

- Inspection to Work Class ROV's
- Ancillary Equipment to Include:
  - Cameras
  - Lights
  - Actuators
  - More

BSS
Baker Subsea Solutions

Inspection ROV's are NOT limited to Inspection scopes!

BSS
Baker Subsea Solutions

ROV Support

Drill Support

IMR

Survey

Inspection

BSS
Baker Subsea Solutions

What determines the ROV requirement?

Scope
- Tooling
- Sensors

Depth
- Shallow Water
- Deep Water

Operational Limits
- Payload
- Horsepower

BSS
Baker Subsea Solutions

What the Spec?

- Depth Rating
- Payload
- Weight
- Thrust
- Telemetry System
- Control System

BMS
Baker Marine Solutions
Providing Solutions for the Marine & Offshore Industries

Evolution of the Inspection Class ROV

- Manipulators
- CP Probe
- UT Probe
- HD Camera
- Sonar
- Sub Bottom Profiler
- Pipe Tracking
- Photogrammetry
- Auto Functions
- Station Keeping
- Fiber
- Bandwidth

BMS Baker Marine Solutions | Providing Solutions for the Marine & Offshore Industries

Evolution of the Inspection Class ROV

Top Hat TMS

- Increase Depth
- Lower Risk Profile - Tether Management
- Lower Risk Profile - Vessel Operations

BMS Baker Marine Solutions | Providing Solutions for the Marine & Offshore Industries

Why use an Inspection ROV?

- Lower Vessel Cost
- Decreased Mobilization Time and Cost
- Maintain Deck Space

BMS Baker Marine Solutions | Providing Solutions for the Marine & Offshore Industries

Timeline – Equipment Growth

Q1 2024

- Hoytek Myra ROV
- 50hp – 70hp ROV
- Launch and Recovery System
- PAMS Control Vans

Q2 2024

- Hoytek Lyra S – Light Work Class ROV

BMS Baker Marine Solutions | Providing Solutions for the Marine & Offshore Industries

Dr. Bertan TEZCAN

HOYTEK
hoytek.com.tr

- EFFICIENCY, RESILIENCE AND TOOLING:
- THE RISING OF HYBRID ROVS IN HEAVY OFFSHORE OPERATIONS

Brief History & Background

HOYTEK

Founded in 2015

Team of 25 people

Delivered more than 30 ROVs

3 Drill Support WROVs with Turkish Petroleum over 25,000 hours each with no downtime in the last 3 years

14 Observation Class Demre ROVs delivered to Police, Coast Guard

1 light work class ROV for archeological research

Lego type structure: mini to WROVs use same electronic components and subsystems

Durable design, good service

HYBRID ROVS

- A **Hybrid** ROV is a relatively new type of a remotely underwater vehicle
  - utilizes a full electric thrust system and a hydraulic tooling system at the same time
  - Each electric thruster >200kgf
  - HPU enables use of legacy tooling equipment

HYDRAULIC ROVS

A Hydraulic ROV is a type of a remotely underwater vehicle which utilizes a full hydraulic system with hydraulic thrusters and hydraulic tooling

- ❖ Higher Power Density
- ❖ Low Initial Costs
- ❖ Oil leak can not be avoided
  - ❖ Water in the system – whole system needs to be flushed
- ❖ Costly and messy maintenance
- ❖ HPU single point of failure

PROS OF HYBRID SYSTEMS

- Simplicity
  - Less Complex Parts – All thrusters and subsystems are decoupled
  - Easy and Quick Maintenance - Very little oil in thrusters, pump oil into compensators when oil level is low
  - HPU runs much less frequently – less maintenance and longer lifetime
- 50% More Efficient
  - Does not overheat in shallow or in warm waters
- 25-30% More Compact
- Up to 50% Lighter – Compact LARS and TMS possible
- Eco Friendly
  - Much Less Fluid Leaks
  - Quieter & Less CO_2 Emissions
- Better Data From Controllers
  - Motor, Diagnostic Data
  - Feedback – Very smooth ROV operation
- Cheaper in the Long Run, much lower Lifetime Cost

HEYTEK LYRA: HEAVY WORK CLASS HYBRID ROV

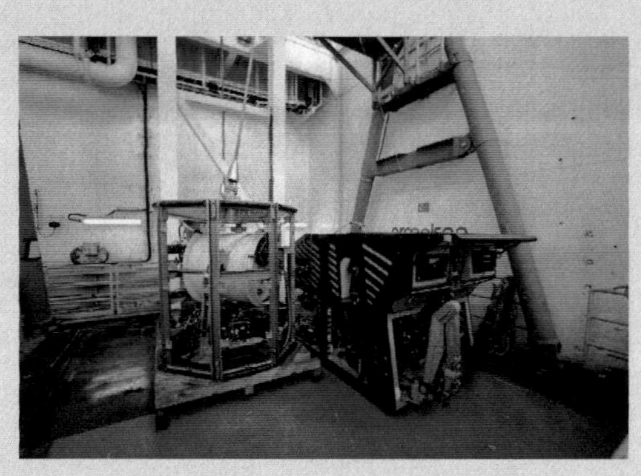

Lyra is designed to be used in operations requiring power and precision such as drilling support, pipeline inspection and intervention, scientific sample collection, exploration, search and rescue, and underwater construction.

- 3000m Depth Rating (6000m optional)
- Schilling T4 ve RigMaster Manipulator Support
- Heavy Duty Tooling Support
- Weight: 3800kg
- Dimensions: 2.8 x 1.8 x 1.7m (LxWxH)
- Standard Payload Capacity: 200kg (After manipulators and HPUs installed)
- Forward Thrust: 800kgf
- Lateral Thrust: 700kgf
- Vertical Thrust: 700kgf
- Fully Customizable

3.000m Depth Rated WROV Kaşif

HOYTEK

7

HOYTEK LYRA: HEAVY WORK CLASS HYBRID ROV: Our Experience

- Field Proven!
- 25,000 Hours of 7/24 Operation
  - 0 hours of Downtime – Much faster maintenance cycle
- Strongest in its Class
- Completed Every Single Drill Support Operation
  - Galaxy tool
  - BOP shear ram closing wrt API 53 standard
- Reliable and Easy Maintenance
  - No single thruster failure in last 4 years with 3 WROW at work
- HPU is not needed to work all the time
  - Much less HPU maintenance compared to Hydraulic ROVs

KAŞİF ROV Operation

HOYTEK

Kaşif - 1

**Fatih
Drill Ship**

Kaşif - 2

**Kanuni
Drill Ship**

Kaşif - 3

**Abdülhamit Han
Drill Ship**

9

KAŞİF-3

HOYTEK

10

Drill Operations

11

HOYTEK LYRA: HEAVY WORK CLASS HYBRID ROV

Partnership with Baker Subsea Solutions

HOYTEK

THANK YOU!

www.hoytek.com.tr

Advancing Underwater Navigation with Dead Reckoning and ROV GPS

Answering the **"Where"** in Who, What, Where, When, Why, How

Presented by: Serena Brown – Account Executive Maritime industries

DEEP TREKKER

WHO: Our Story/ History

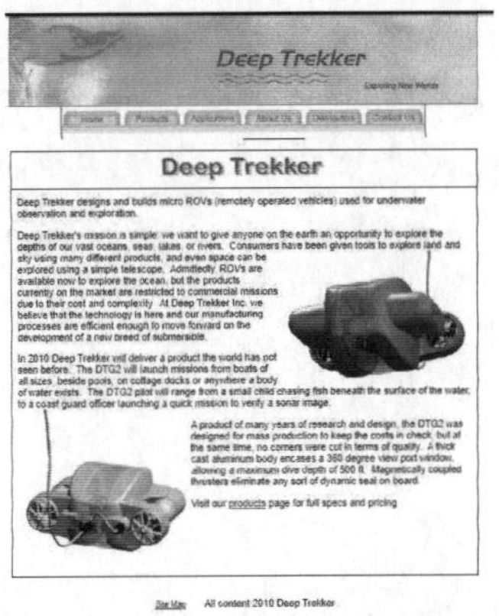

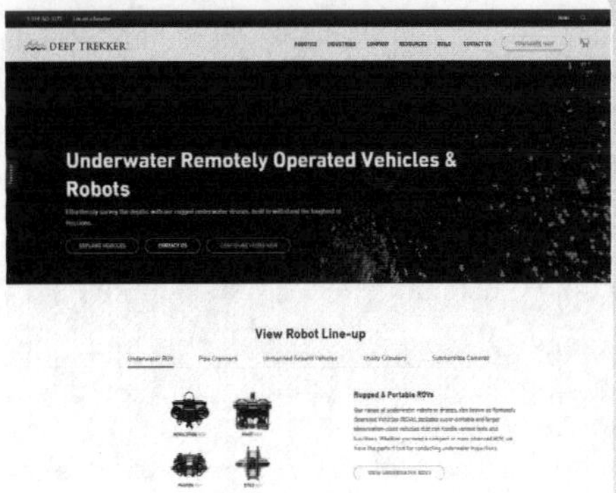

2010 – 2024

WHO: Trusted By...

WHAT: Industries We Serve

Defense	Public Safety	Aquaculture	Marine Science
Environmental	Shipping	Hydroelectric	Offshore
Commercial Diving	Civil Infrastructure	Wastewater/ Stormwater	Industrials / Petrochemical
Water Utilities	Nuclear	Wind	Mining

Remotely Operated Vehicles

PHOTON ROV
PIVOT ROV
REVOLUTION ROV
Powered by Deep Trekker BRIDGE Technology

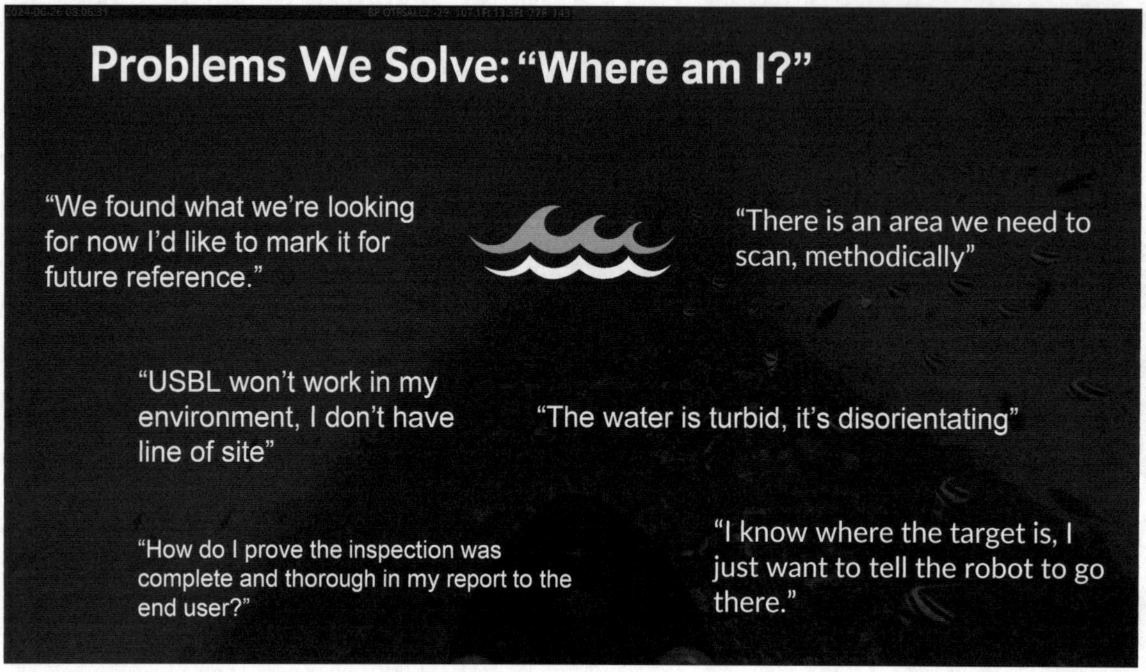
Problems We Solve: “Where am I?”
“We found what we’re looking for now I’d like to mark it for future reference.”
“There is an area we need to scan, methodically”
“USBL won’t work in my environment, I don’t have line of site”
“The water is turbid, it’s disorientating”
“How do I prove the inspection was complete and thorough in my report to the end user?”
“I know where the target is, I just want to tell the robot to go there.”

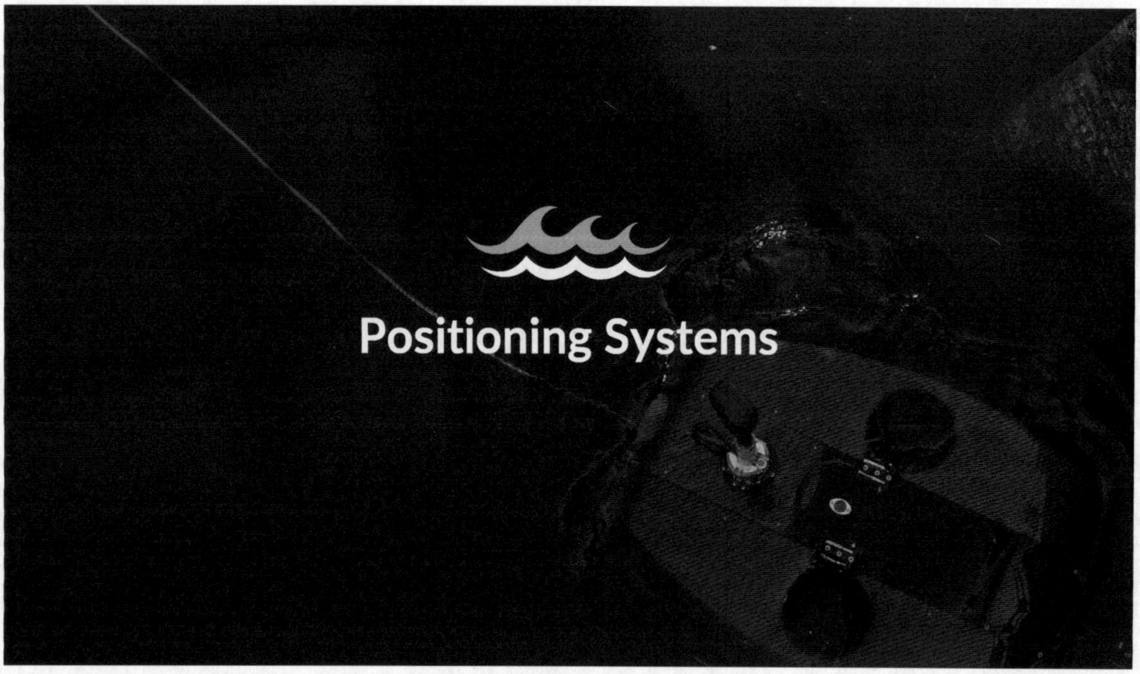
Positioning Systems

Positioning Setups and How they tie together

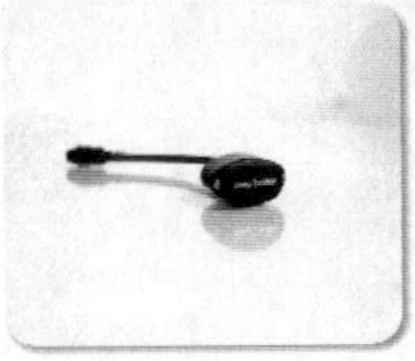

Sensor Pod

Sensor Pod+

A125 DVL

A50 DVL

SeaTrac USBL

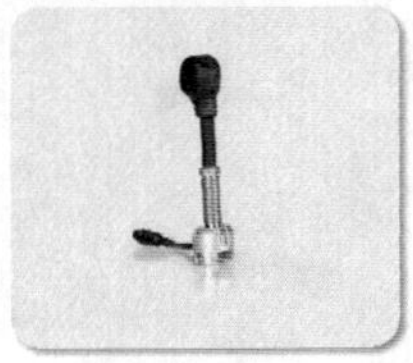

ROV GPS

How does USBL work?

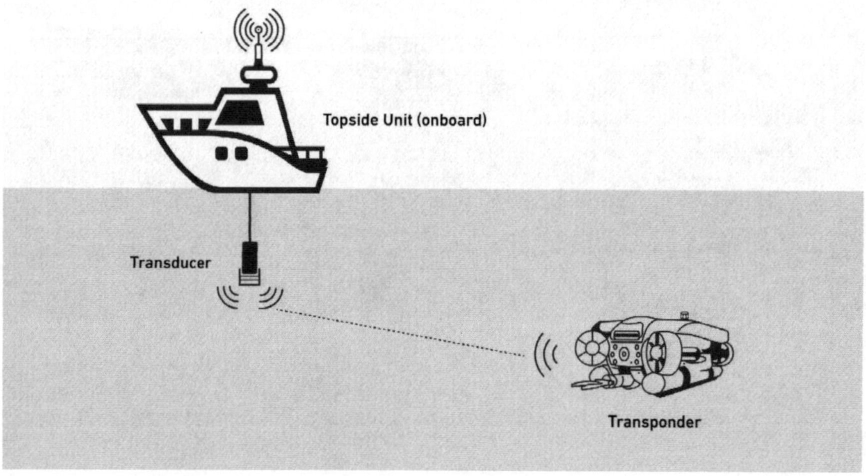

Dead Reckoning at a Glance

- Autonomous Operations
- Rapid Single Operator Deployment
- Route Planning
- GPS Waypoint Data Logging
- Automatic Waypoint Navigation
- ~30 min Run Time between Pings
- No External Position Sensors Required
- Quick Reset

How does GPS work?

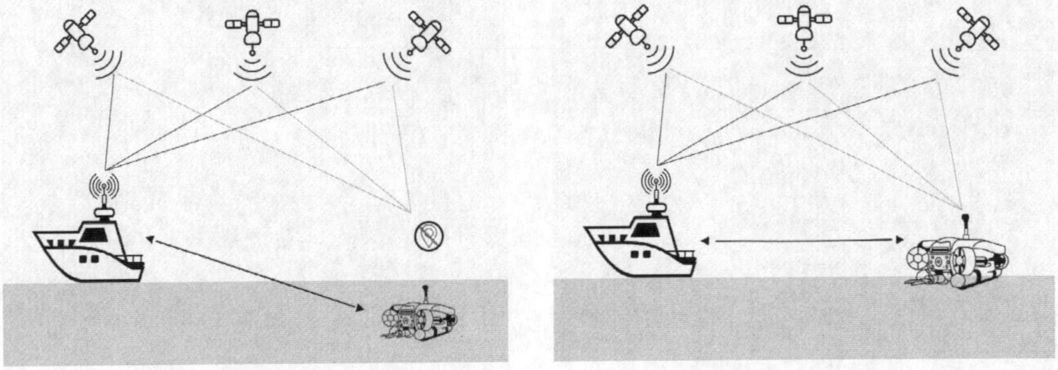

No GPS Underwater

GPS Signal Above Water

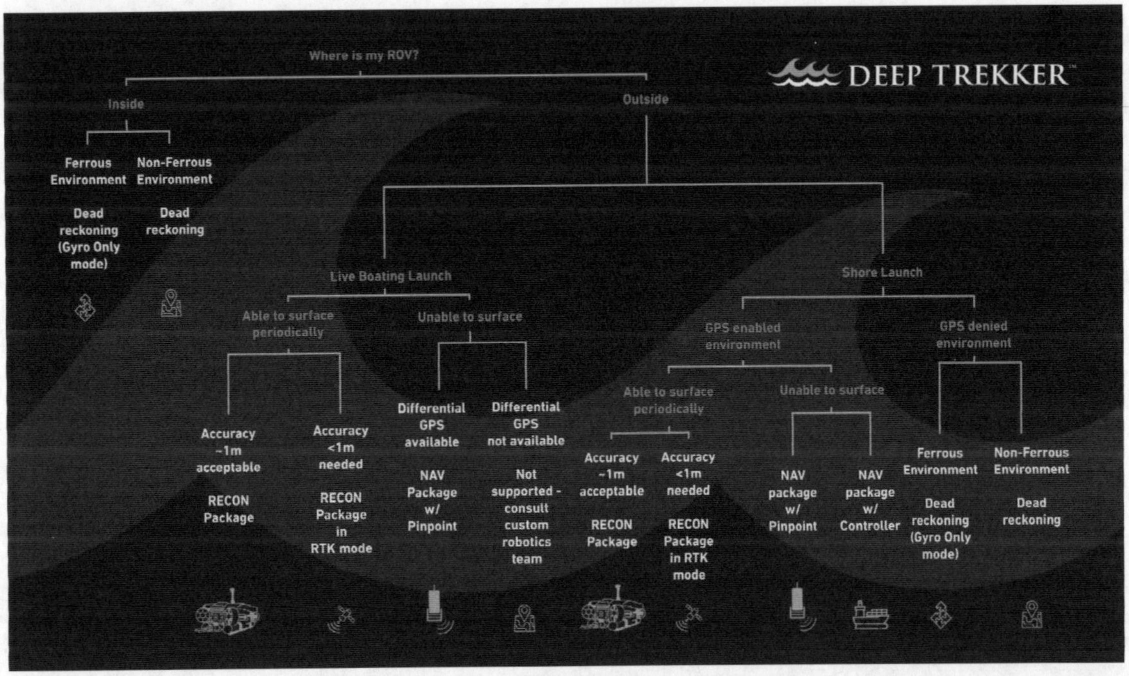

DEEP TREKKER™

imenco
SMART SOLUTIONS

ROV Obstacle Alert

'Parking sensors' for your ROV

imenco.com

imenco
SMART SOLUTIONS

Full coverage to detect obstacles around the ROV

Avoid undue costs due to collisions

imenco.com

imenco
SMART SOLUTIONS

The problem

- Increased ROV activity
- Turnover / new generation of ROV pilot expertise
- Increased incidents of collisions
  - Repair cost to the:
    - ROV
    - sensors (spares?)
    - payload (replacable?)
    - asset (integrity risk?)
- Potential downtime cost

imenco.com

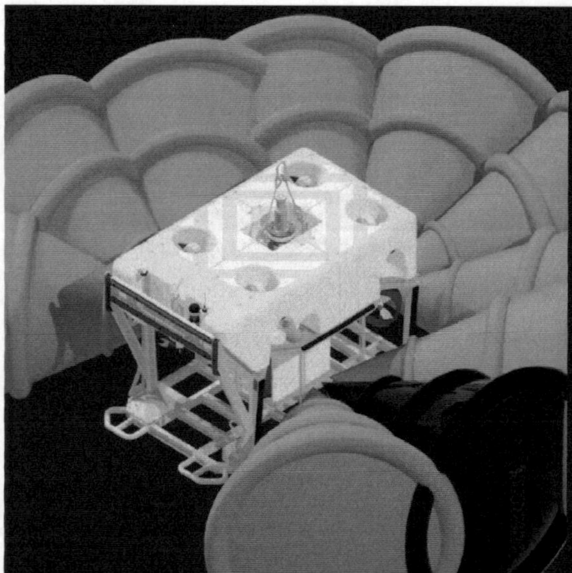

imenco
SMART SOLUTIONS

Client's challenge

- Identify multiple obstacles in proximity to the ROV
- Operate in high turbidity
- Minimize blind spots
- Identify proximity to the TMS
- Minimal burden on the ROV
  - Small form factor
  - Low power
  - Low data bandwidth requirement
- Easy to understand interface, minimal interpretation
- Low cost

imenco.com

imenco
SMART SOLUTIONS

Solution Assessment

- **Scanning sonar** - Challenge
  - ✓ Identify multiple obstacles in proximity to the ROV
  - ✓ Operate in high turbidity
  - ✓ Minimize blind spots
  - ✓ Identify proximity to the TMS
  - ✗ Minimal burden on the ROV
    - Small form factor
    - Low power
    - Low data bandwidth requirement
  - ✗ Easy to understand interface, minimal interpretation
  - ✗ Low cost

- Visual cameras - Challenge
  - ✓ Identify multiple obstacles in proximity to the ROV
  - ✗ Operate in high turbidity
  - ✓ Minimize blind spots
  - ✓ Identify proximity to the TMS
  - ✗ Minimal burden on the ROV
    - Small form factor
    - Low power
    - Low data bandwidth requirement
  - ✗ Easy to understand interface, minimal interpretation
  - ✗ Low cost

imenco.com

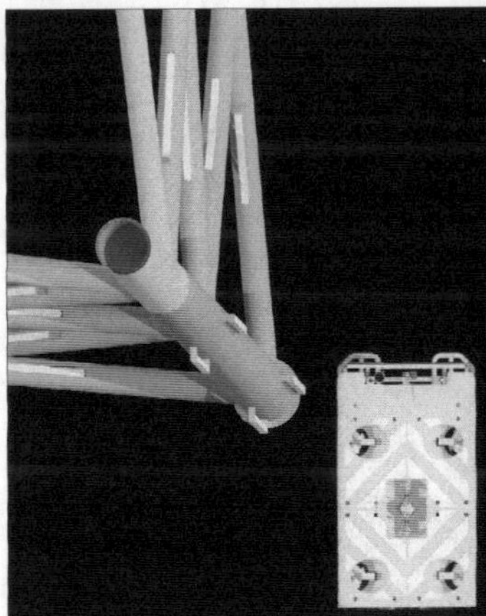

imenco
SMART SOLUTIONS

Our Imenco Smart Solution

- Array of acoustic sensors / echosounders
  - 4x on the rear
  - 3x on both sides
  - 2x on the front
  - 1x above
  - 1x below
- 45 degree detection sectors
- 0.5 – 100 m measurement range
  - Expect 3-5 m maximum range of interest
- 2.5 mm range resolution
- 1 Hz update rate for the whole array
- Graphical display, or API to interface to existing nav displays

imenco.com

imenco
SMART SOLUTIONS

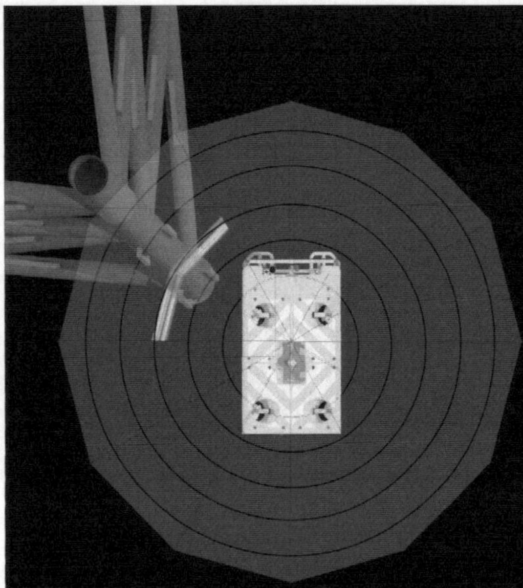

Our Imenco Smart Solution

- Array of acoustic sensors / echosounders
  - 4x on the rear
  - 3x on both sides
  - 2x on the front
  - 1x above
  - 1x below
- 45 degree detection sectors
- 0.5 – 100 m measurement range
  - Expect 3-5 m maximum range of interest
- 2.5 mm range resolution
- 1 Hz update rate for the whole array
- Graphical display, or API to interface to existing nav displays

imenco.com

imenco
SMART SOLUTIONS

Components

- 1x Subsea Router
  - Depth rated: 3500 m
  - Robust electronics
  - Standard connectors
  - Small form factor for flexibility of placement
  - 251 x Ø90 mm
- Mounting Brackets
  - Simple water jet stainless steel plate maintaining relative sensor alignment
  - Standard off-the-shelf injection molded pipe clamps
  - Durable, low-cost solution
- Transducers
  - Depth rated: 3500 m
  - Low Power
  - Frequency: 115 kHz
    - Unlikely to interfere with other acoustics
  - Small form factor for flexibility of placement
  - 119 x Ø55 mm

imenco.com

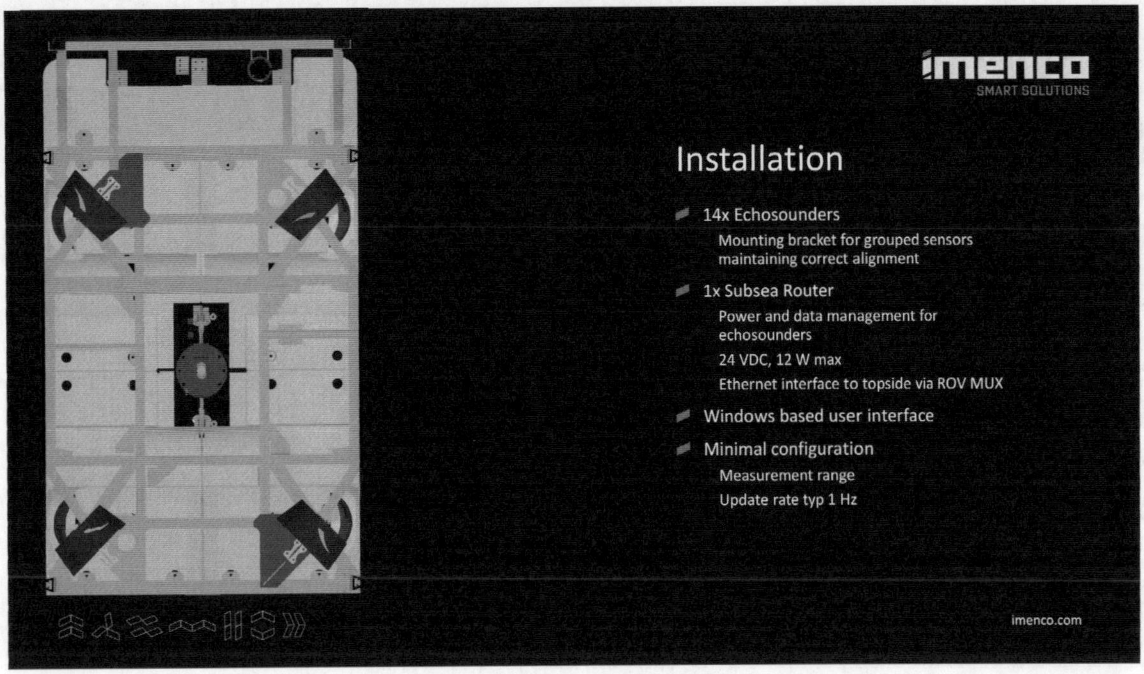
imenco
SMART SOLUTIONS
Installation
14x Echosounders
Mounting bracket for grouped sensors maintaining correct alignment
1x Subsea Router
Power and data management for echosounders
24 VDC, 12 W max
Ethernet interface to topside via ROV MUX
Windows based user interface
Minimal configuration
Measurement range
Update rate typ 1 Hz
imenco.com

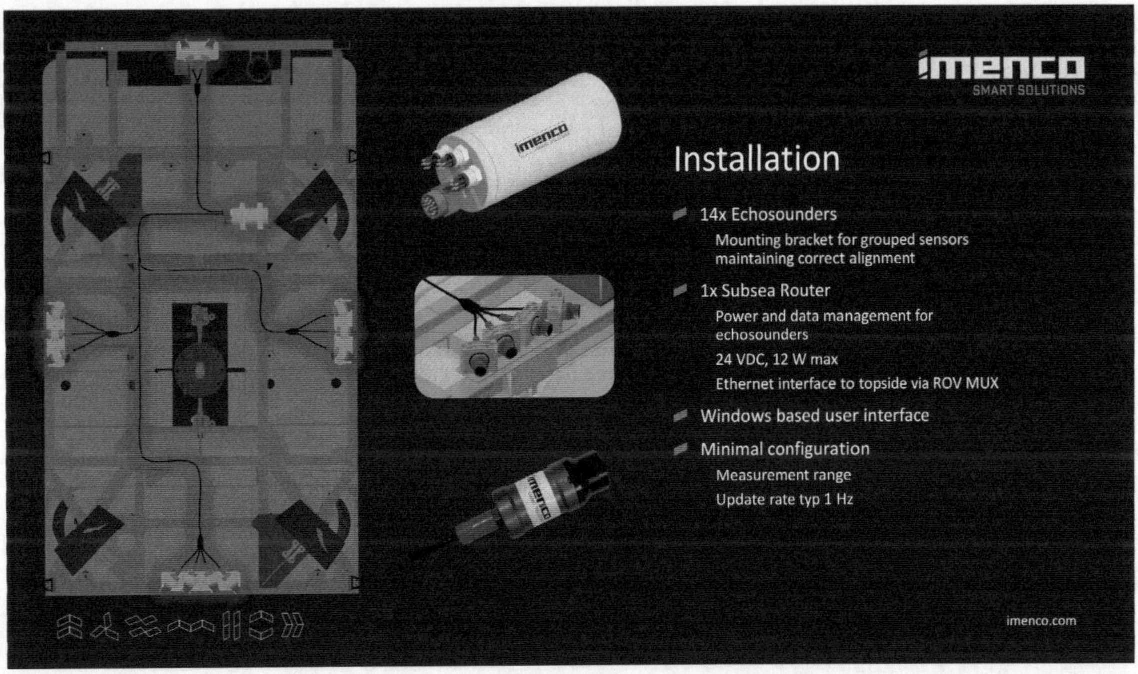
imenco
SMART SOLUTIONS
Installation
14x Echosounders
Mounting bracket for grouped sensors maintaining correct alignment
1x Subsea Router
Power and data management for echosounders
24 VDC, 12 W max
Ethernet interface to topside via ROV MUX
Windows based user interface
Minimal configuration
Measurement range
Update rate typ 1 Hz
imenco.com

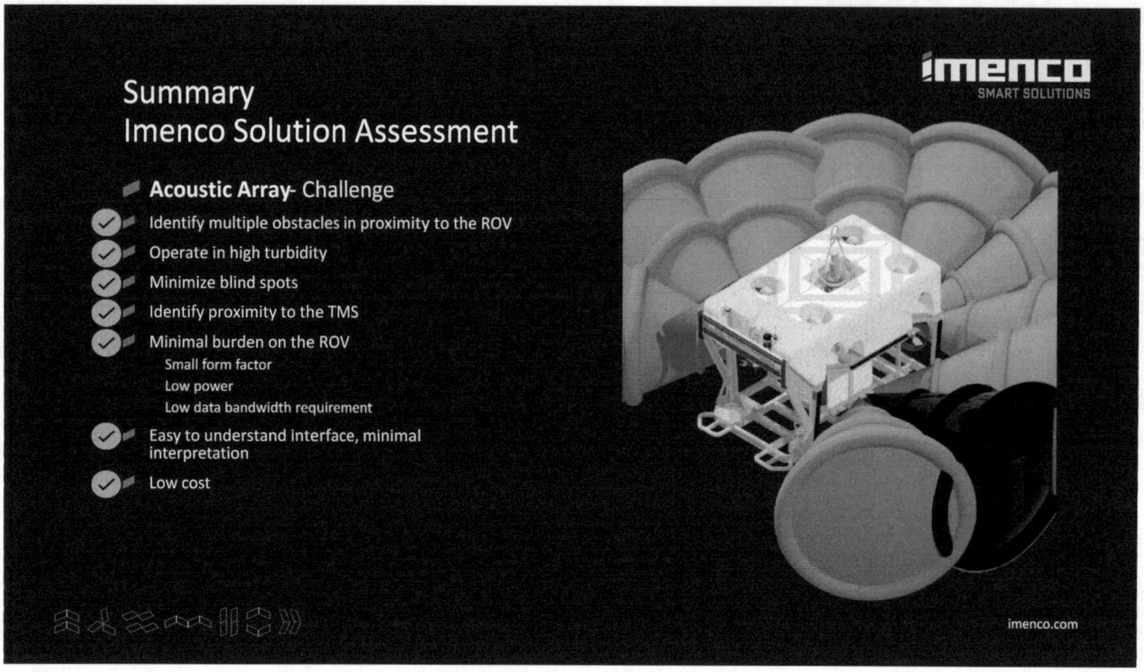
imenco
SMART SOLUTIONS
Summary
Imenco Solution Assessment
Acoustic Array- Challenge
Identify multiple obstacles in proximity to the ROV
Operate in high turbidity
Minimize blind spots
Identify proximity to the TMS
Minimal burden on the ROV
Small form factor
Low power
Low data bandwidth requirement
Easy to understand interface, minimal interpretation
Low cost
imenco.com

imenco
SMART SOLUTIONS
Andrew Connelly
Product Line Manager – Acoustics
Mobile: +44 (0) 7741 29 3666
E-mail: Andrew.Connelly@imenco.com
imenco.com

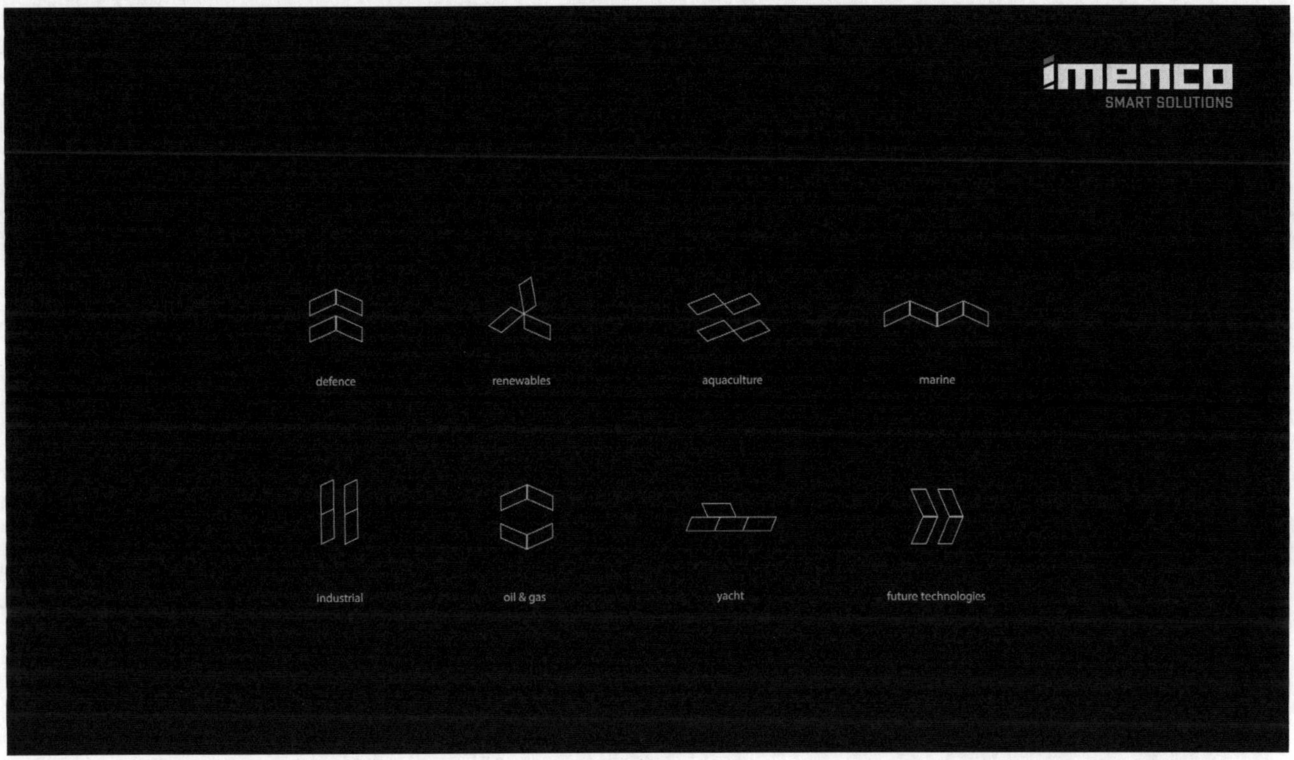
imenco
SMART SOLUTIONS
defence
renewables
aquaculture
marine
industrial
oil & gas
yacht
future technologies

AUTHOR INDEX